Science Networks · Historical Studies
Volume 7

Edited by Erwin Hiebert and Hans Wussing

Editorial Board:

S. M. R. Ansari, Aligarh
D. Barkan, Pasadena
H. J. M. Bos, Utrecht
U. Bottazzini, Bologna
J. Z. Buchwald, Toronto
S. S. Demidov, Moskva
J. Dhombres, Nantes
J. Dobrzycki, Warszawa
Fan Dainian, Beijing
E. A. Fellmann, Basel
M. Folkerts, München
P. Galison, Stanford
I. Grattan-Guinness, Bengeo
J. Gray, Milton Keynes

R. Halleux, Liège
S. Hildebrandt, Bonn
E. Knobloch, Berlin
Ch. Meinel, Regensburg
G. K. Mikhailov, Moskva
S. Nakayama, Tokyo
L. Nový, Praha
D. Pingree, Providence
W. Purkert, Leipzig
J. S. Rigden, New York
D. Rowe, Pleasantville
A. I. Sabra, Cambridge
R. H. Stuewer, Minneapolis
V. P. Vizgin, Moskva

Birkhäuser Verlag
Basel · Boston · Berlin

Fyodor A. Medvedev

Scenes from the History of Real Functions

Translated from the Russian by Roger Cooke

1991

Birkhäuser Verlag
Basel · Boston · Berlin

Author's address

Professor Dr. F. A. Medvedev
Institute for History
of Science and Technology
Staropanskii Pereulok, d. 1/5
103012 Moskva, USSR

Originally published as: Ocherki istorii teorii funktsii deistvitel'nogo peremennogo,
Moskva: Nauka 1975

Deutsche Bibliothek Cataloging-in-Publication Data

Medvedev, Fedor A.:
Scenes from the history of real functions / Fyodor A.
Medvedev. Transl. from the Russian by Roger Cooke. – Basel ;
Boston ; Berlin : Birkhäuser, 1991
  (Science networks ; Vol. 7)
  Einheitssacht.: Očerki istorii teorii funkcij dejstvitel'nogo
  peremennogo <engl.>
  ISBN 3-7643-2572-0
NE: GT

This work is subject to copyright. All rights are reserved, whether the whole or part
of the material is concerned, specifically those of translation, reprinting, re-use of
illustrations, broadcasting, reproduction by photocopying machine or similar means,
and storage in data banks. Under § 54 of the German Copyright Law where copies are
made for other than private use a fee is payable to »Verwertungsgesellschaft Wort«, Munich.

© 1991 for the English edition: Birkhäuser Verlag Basel
Printed from the translator's camera-ready manuscript on acid-free paper in Germany
ISBN 3-7643-2572-0
ISBN 0-8176-2572-0

Translator's Note

The Russian original of the present work, bearing the title Очерки Истории Теории Функций Действительного Переменного (Essays on the History of the Theory of Functions of a Real Variable), was published in 1975. The present translation incorporates some references to historico-mathematical work that has appeared since that time along with corrections of the misprints that were detected in the original. The new references and the corrections were both provided to the translator by Prof. Medvedev. These are the only elements added in the present work, which is otherwise a complete translation of the original. The following points, however, should be noted:

1. For the sake of euphony and brevity the English title is not a translation of the Russian, but has been chosen to convey the general contents of the book as described by Prof. Medvedev in the preface—a selection of topics from the history of the theory of functions of a real variable.

2. The original work generally referred to Russian translations of mathematical works whenever such an edition existed. These Russian translations have in every case been replaced by their originals in the bibliography; and whenever the originals were available to the translator, page citations to them in the text have been amended accordingly. In the two cases where the originals were not available—the references to Cavalieri's *Exercitationes* in Section 2 of Chapter 4 and the references to Cauchy's *Calcul infinitésimale* in Section 5 of Chapter 4—the Russian page references were used in the hope that they may provide at least a clue as to the location referred to.

3. Prof. Medvedev's Russian translations of quotations from English works have been replaced by the original English, except for the quotation from H. G. Wells in the preface, which the translator was unable to locate.

Contents

Preface 9

Chapter 1 The place of the theory of functions of a real variable among the mathematical disciplines

1.1 The subject matter of the theory of functions 11
1.2 Three periods in the development of
 the theory of functions 13
1.3 The theory of functions and classical analysis 15
1.4 The theory of functions and functional analysis 18
1.5 The theory of functions and other
 mathematical disciplines 22

Chapter 2 The history of the concept of a function

2.1 Some textbook definitions of the concept of a function 25
2.2 The concept of a function in ancient times and in the
 Middle Ages 29
2.3 The seventeenth-century origins of the concept
 of a function 32
2.4 Some particular approaches to the concept of a function
 in the seventeenth century 35
2.5 The Eulerian period in the development of the
 concept of a function 40
2.6 Euler's contemporaries and heirs 44
2.7 The arbitrariness in a functional correspondence 50
2.8 The Lobachevskii-Dirichlet definition 55
2.9 The extension and enrichment of the concept of a
 function in the nineteenth century 63
2.10 The definition of a function according to Dedekind 67
2.11 Approaches to the concept of a function
 from mathematical logic 70
2.12 Set functions 73
2.13 Some other functional correspondences 77

Chapter 3 Sequences of functions. Various kinds of convergence

3.1 The analytic representation of a function 82

3.2	Simple uniform convergence	85
3.3	Generalized uniform convergence	99
3.4	Arzelà quasiuniform convergence	104
3.5	Convergence almost everywhere	110
3.6	Convergence in measure	120
3.7	Convergence in square-mean. Harnack's unsuccessful approach	128
3.8	Square-mean convergence. The work of Fischer and certain related investigations	137
3.9	Strong and weak convergence	145
3.10	The Baire classification	157

Chapter 4 The derivative and the integral in their historical connection

4.1	Some general observations	168
4.2	Integral and differential methods up to the first half of the seventeenth century	170
4.3	The analysis of Newton and Leibniz	177
4.4	The groundwork for separating the concepts of derivative and integral	181
4.5	The separation of differentiation and integration	187
4.6	The Radon-Nikodým theorem	198
4.7	The relation between differentiation and integration in the works of Kolmogorov	202
4.8	The relation between differentiation and integration in the works of Carathéodory	205
4.9	A few more general remarks	208

Chapter 5 Nondifferentiable continuous functions

5.1	Some introductory remarks	212
5.2	Ampère's theorem	214
5.3	Doubts and refutations	219
5.4	Classes of nondifferentiable functions	224
5.5	The relative "smallness" of the set of differentiable functions	228

Bibliography	235
Index of names	260

Preface

To attempt to compile a relatively complete bibliography of the theory of functions of a real variable with the requisite bibliographical data, to enumerate the names of the mathematicians who have studied this subject, exhibit their fundamental results, and also include the most essential biographical data about them, to conduct an inventory of the concepts and methods that have been and continue to be applied in the theory of functions of a real variable... in short, to carry out any one of these projects with appropriate completeness would require a separate book involving a corresponding amount of work. For that reason the word *essays* occurs in the title of the present work, allowing some freedom in the selection of material.

In justification of this selection, it is reasonable to try to characterize to some degree the subject to whose history these essays are devoted. The truth of the matter is that this is a hopeless enterprise if one requires such a characterization to be exhaustively complete and concise. No living subject can be given a final definition without provoking some objections, usually serious ones. But if we make no such claims, a characterization is possible; and if the first essay of the present book appears unconvincing to anyone, the reason is the personal fault of the author, and not the objective necessity of the attempt.

It can hardly raise objections in a set of essays on the history of the theory of functions[1] to designate a special section devoted to the history of the central concept of this subject. One may rather complain that it is too brief and incomplete, or perhaps unsatisfactory. But here again the reproach lies against the author rather than the actual state of affairs.

Clearly a section on sequences of functions is equally legitimate. The various kinds of convergence of functional sequences play too prominent a role in the theory of functions to be denied their share of attention, even in the most modest *essays*.

The connection between the operations of differentiation and integration led to the rise of classical analysis. Classical analysis in turn is responsible in large degree for the development of the theory of functions. And if a breach between the concepts of derivative and integral is characteristic of the latter, nevertheless the attempt to establish a connection of some kind between them

---

[1] Here and below the words *theory of functions* are used for brevity to denote *theory of functions of a real variable*.

was one of the most important stimuli for the advancement of mathematical thought in this area. To trace the vicissitudes of the relationship between the two fundamental infinitesimal operations is the task of the fourth essay.

The study of nondifferentiable functions, begun in the 1870's and continuing down to the present, led to an essential re-examination of certain entrenched conceptions. In the concluding part of the present work an attempt is made to describe particular aspects of the history of this problem.

The essay/chapters just named by no means cover the subject of the history of the theory of functions, even if one imagines them written with exhaustive completeness: Enormous gulfs remain almost entirely beyond the scope of this book, such as the constructive theory of functions in the sense of Chebyshev-Bernshtein, the theory of quasianalytic and almost-periodic functions, and much more. The theories of differentiation and integration could be the subject of huge separate books, and in the fourth essay only particular moments of the history of these operations are presented. The same should be said about the relationship between the history of the theory of series and the third essay. In short, there are more than enough gaps to make the present book on the history of the theory of functions a good exercise book for the critically inclined reader. Perhaps what is written in this book will provoke even more critical comments. To this one could reply jokingly with the following words of H. G. Wells: "Anyone who does not like this book is free to choose another one or to write a better one according to his own taste and measure."

Now a few words about the bibliography at the end of the book. First of all, each article or book mentioned there was studied by the author to some extent. Furthermore, we could easily have made the bibliography several times larger than it is while preserving the quality mentioned in the preceding statement, and only the desire to reduce the volume of the book forced us to make a selection, whose correctness it is left to the reader to decide. The year of first publication of a work is noted at the end of each entry. In citations only the most fundamental work is cited: preliminary communications on the same theme are not mentioned when it is not necessary to do so.

CHOICE
100 Riverview Center
Middletown, CT 06457
Tel. (203) 347-6933   FAX (203) 346-8586

Using a computer or word processor printer for your review

Your reviews need not be printed on *CHOICE* forms if these do not work easily with your printer. You may send a separate sheet along with the *CHOICE* form, which we need for record keeping, but **Please** observe the following:

— Double space
— Do not paragraph
— Do not use an all-italic or all-cap typeface

To keep your review to our 175-word limit

— Set a 70-character line count

If you use 12 pitch (elite) review length is 17 lines; with 10 pitch (pica) review length is 19 lines. Please do not use any typeface smaller than 12 pitch.

You may send the review printed on a separate sheet. We will tape it to the *CHOICE* review form to give our data entry staff only one piece of paper to handle. If you follow the guidelines above, especially as to line length, your review will fit the form.

Late Arrivals: Books and/or Reviews

CHOICE schedules 4 weeks for review completion, setting due dates 5 weeks from the day books are sent. If your book is delayed in arriving, please let us know by letter or phone if the delay must make your review more than 2 weeks late.

About 2 weeks after the original due date we will query you if your review has not arrived. Please ALWAYS check and return any query form in case the review may have been lost in the mail.

Chapter 1
The place of the theory of functions of a real variable among the mathematical disciplines

1.1 The subject matter of the theory of functions

Examining the monographs and textbooks on the theory of functions, one may notice that, as a rule, they do not give any comparatively brief definition of the subject matter of this mathematical discipline, as is done, for example, for topology—the study of the properties of topological spaces that are invariant under homeomorphisms—or algebra—the study of operations on elements of sets of arbitrary nature. The definition of the theory of functions occasionally encountered, as the study of the general properties of functions, is essentially a tautology.

The subject matter of the theory of functions is most often characterized by enumerating the problems that are studied in it. Such an enumeration is either carried out explicitly (usually in the introductions to books) or is implicit in the table of contents. We now give several examples of what we are talking about.

In the first treatise on the theory of functions, Dini's book *Fondamenti per la teorica delle funzioni di variabili reali* [4], the list of chapters is as follows: "Irrational numbers"[1]; "Groups of numbers and points, their upper and lower limits"[2]; "The concept of a limit. Infinitely small and infinitely large quantities"; "The concept of a function. Continuity and discontinuity"; "Functions continuous on a given interval"; "Functions having an infinite number of discontinuities"; "Derivatives"; "Theorems on series"; "The principle of condensation of singularities"; "Functions having a definite finite derivative nowhere"; "Other considerations relating especially to the existence of the derivative of finite and continuous functions"; "Definite integrals."

In the introduction to the book of Hobson *The Theory of Functions of a Real Variable and the Theory of Fourier Series* (1907), in particular it is stated that, "The theory of functions of a real variable, as developed during

---

[1] Dedekind's theory of real numbers [2] is meant.

[2] This chapter contains an exposition of the earliest results on the theory of point sets obtained by mathematicians up to 1872. Dini calls the least upper bound and greatest lower bound the upper limit and lower limit respectively.

the last few decades, is a body of doctrine resting, first upon the conception of the arithmetic continuum which forms the field of the variable, and which includes a precise arithmetic theory of the nature of a limit, and secondly, upon a definite conception of the nature of the functional relation. The procedure of the theory consists largely in the development, based upon precise definitions, of a classification of functions, according as they possess, or do not possess, certain peculiarities, such as continuity, differentiability, etc., throughout the domain of the variable or at points forming a selected set contained in that domain. The detailed consequences of the presence, or of the absence, of such peculiarities are then traced out, and are applied for the purpose of obtaining conditions for the validity of the operations of Mathematical Analysis" [2, p. V].

If we supplement this rather disjointed statement of the contents of of Hobson's book, we can say that he depicts the subject matter of the theory of functions in approximately the same form as Dini, with a suitable enlargement and deepening (and sometimes, however, impoverishment) of the results obtained in set theory and the theory of functions after the appearance of Dini's book, especially in the theory of trigonometric series, which was not considered at all in Dini's book.

Finally, the contents of one of the modern textbooks, that of Natanson, *Theory of Functions of a Real Variable* [1], are presented as follows in the table of contents. Five of the eighteen chapters are devoted to set theory, eight to various types and classes of functions, three to integration theory, one to singular integrals, trigonometric series, and convex functions, and one to elementary functional analysis.[3]

If we attempt to characterize the subject matter of the theory of functions by enumerating its fundamental divisions, the following picture emerges.

First of all, as an essential condition for the exposition of the subject, it is considered necessary to preface the circle of set-theoretic ideas, which continues to enlarge with time, beginning with the modest results given by Dini and ending with the extensive results of Natanson. In the latter book these results are found not only in the chapters listed above, but also within chapters assigned to the other groups. This situation is not at all coincidental, but is inherent in the very nature of things: the theory of functions of a real variable was constructed together with set theory. They have always been, and continue to be, closely related to each other; and if one considers only the theory of point sets in Euclidean spaces and the study of functions on such

---

[3] Such a division of chapters into groups is a convention, since, for example, in Chapter VI, titled "Integrable functions," the subject is mostly integration, while in Chapter VII, "Functions of bounded variation. The Stieltjes integral," the subject is an important class of functions.

sets, they are often simply identified.

Furthermore the study of types and classes of functions naturally turns out to be a most important component part of the theory of functions—differentiable and nondifferentiable, continuous, monotonic, integrable, measurable (to a lesser degree, nonmeasurable), set functions, and the like.

An exceptionally large division of the theory of functions, which really can be distinguished as a separate subject, is the theory of series.

Two operations on functions, differentiation and integration, are the basic tools in the study of functions, along with the representation of functions by series. Closely connected with these tools is the concept of passage to the limit—an operation that is a rather inhomogeneous mixture of different types of limiting passages. (In a certain sense the first two operations comprise similar mixtures.)

Such are the main components of the theory of functions if we do not include in it the constructive function theories or the theories of almost-periodic and quasianalytic functions.

1.2 Three periods in the development of the theory of functions

Periodization is always a rather complicated and delicate matter. Recognizing this, we have nevertheless chosen to designate three large periods, which in our view are rather clearly marked. At this point they will not be given a detailed characterization, which would form the subject of a large separate article. Furthermore we do not propose to go into details as to the way in which the question of the origin of the theory of functions has been treated in the literature, a question closely connected with the problem of periodization. Finally, the reader should be warned that the proposed periodization does not allow a clear demarcation of the boundaries of the periods: they sometimes overlap to such an extent that it is difficult to say to which period many of the facts of the history of the theory of functions should be assigned, and the occurrence of such-and-such an event in the history under consideration during one of these time intervals or another may in no way correspond to the general characteristics of the period, since the periods do overlap in many respects.

The following three periods are distinguished in the present set of essays:

I. The period from the appearance of Riemann's paper "Über die Darstellbarkeit der Funktionen durch trigonometrische Reihen" [2] in 1867[4] to the first works of Borel, Baire, and Lebesgue in the period 1895–1902.

---

[4] The date of publication is variously given as 1867 or 1868 in different sources.

14   Chapter 1   The theory of functions among the mathematical disciplines

II. The period from the first works of the above-named French mathematicians to the appearance of the books of Banach (*Théorie des opérations linéaires* [2]), Saks (*Theory of the Integral* [2]), and Kaczmarz and Steinhaus (*Theory of Orthogonal Series* [1]), which were published in the 1930's.

III. From the 1930's to the present.

The first of these can be characterized as a period of tightly interwoven investigations, particularly in the area of classical analysis with applied investigations in the theory of functions. The works of Riemann, Darboux, Du Bois-Reymond, Cantor, Dini, and many other authors date from this period. The basic results were: the introduction and study of the Riemann integral, various derivatives (including the Dini derivates), the beginning of the study of general trigonometric series, and other expansions in orthogonal functions, and the development of point set theory.

The second period can be called the period of comparatively independent development of the theory of functions of a real variable as a separate mathematical discipline based on point set theory in n-dimensional Euclidean space. The basic results are: the introduction and study of the integrals of Lebesgue, Denjoy, and Perron and a variety of other integration processes, especially those of Stieltjes type; more profound studies of differentiation; the development of methods of summing series; the designation of quite broad and important classes of functions, especially set functions. Hundreds of great and lesser scholars from a great many countries participated in the development of this new subject.

The third period is characterized by the close ties of the theory of functions, mainly with functional analysis. The latter had developed under the influence of the former for a time, but by the middle 1930's it had grown so much that it gradually began to subsume the traditional areas of the theory of functions. The theories of differentiation and integration began to be developed primarily within the framework of functional analysis, and the latter penetrated deeply into summation methods and many other traditional areas of investigation.

We now give several examples of the overlap of these periods in both their temporal and their spatial aspects.

The integral in the sense of the Cauchy principal value was introduced in 1823. The representatives of the first stage in the development of the theory of functions, with very few exceptions, were more than cool toward it. In the years 1901–1903 Hardy resurrected this concept, originally so to speak within the framework of classical analysis, and for some time it was ineffective. It was only in 1909, after Hardy connected it with function-theoretic content and applied it in the theory of integral equations that the integral in the sense of the Cauchy principal value became an important tool of investigation in many questions of mathematics. Thus this integral, which had appeared in the prefunctional period, began to be used only in the second and especially in the third period of the development of the theory of functions.

On the other hand, we began the real history of the theory of functions of a real variable with the introduction of the Riemann integral. From 1867 to the end of the nineteenth century a large amount of work was done to provide a foundation for the theory of Riemann integration. Meanwhile up to the beginning of the twentieth century many mathematicians managed "not to notice" this most important tool in the study of mathematical facts, continuing to use the integral in the form of the difference of the values of a primitive function and ignoring the many deficiencies of such an approach.

As many examples as desired can be given of works that would belong to the second period according to its characteristics but were carried out during the third period. They continue down to the present, and many of them will be pointed out below. But they are not the ones that determine the shape of the theory of functions after the 1930's.

Finally, many works in which the ideas of functional analysis were applied in the solution of function-theoretic problems appeared long before the beginning of the third period. Examples will be given below.

Thus the indicated periodization of the history of the theory of functions is rather hypothetical. Nevertheless we consider that it reflects the state of affairs with reasonable accuracy, and we shall use it in the following exposition.

1.3 The theory of functions and classical analysis

A comparison of the theory of functions with classical analysis shows that all the branches of the former enumerated above (except the introduction of set theory) were also branches of ordinary analysis. The latter also involved the study of various kinds of functions, series of them, the operations of differentiation and integration on them, and various kinds of limiting passages; but in the theory of functions all this is raised to a higher level of abstraction.

If we exclude from classical analysis the theory of differential equations and the theory of functions of a complex variable, which separated off as enormous independent areas of mathematical knowledge, we can say that the theory of functions of a real variable is simply a larger, deeper, and more general mathematical analysis.[5] In this respect the second edition of Jordan's *Cours d'analyse* [3] is very typical. On the one hand it is indeed a treatise on classical analysis, including all of its main branches, and the title of the book

---
[5] However, particular parts of the theory of differential equations are sometimes included in a course in the theory of functions. Cf., for example, Bourbaki [2].

corresponds perfectly to its content. On the other hand it is fully justified to call this book of Jordan's a treatise on the theory of functions. The ideas and methods of the two mathematical disciplines are so tightly interwoven in it that it is practically impossible to distinguish them. The same can be said about all the editions of Vallée-Poussin's *Cours d'analyse infinitesimale* [1]: it could be called with even more justification a textbook on the theory of functions. The two books just named—and their number could be greatly increased—illustrate very well the close connection between analysis and the theory of functions. It is nevertheless an interesting historical circumstance that in 1878 Dini had already distinguished the theory of functions as an independent branch of mathematics, correctly forecast its content, and given it a name, and that mathematicians had basically accepted his approach.

The close connection between analysis and the theory of functions, and the sometimes undetectable transitions from one to the other raise the question of the relationship of these two mathematical disciplines. Are they really separate subjects, and if so, how do they differ from one another?

The first difference that appears is the absence of explicit set-theoretic ideas in classical analysis and the massive application of them in the theory of functions. The point of departure in analysis—the function—is always defined on some domain that is not divided into parts—an interval of the line, a domain in the plane, a volume in space. In contrast, in the theory of functions it is generally defined on some set of points, and its properties depend essentially on the character of the set on which it is defined[6]; it is therefore natural that here the study of functions is preceded by the study of sets. Moreover such a preliminary study is necessary for the additional reason that the solution of many problems of the theory of functions depends on the solution of set-theoretic problems. We shall give just one example. To answer the question whether given function is integrable in classical analysis it was necessary either simply to calculate its integral, or to show that the function belongs to a certain class of functions for which the concept of integral is defined; and that class itself is introduced by a method making no external appeal to set theory. In contrast, to answer the question of the integrability of a function in the Riemann sense, it is necessary and sufficient to know whether the set of its discontinuities is of measure zero.

Such is the state of affairs in the majority of questions of the theory of

---

[6] In general such a dependence holds in analysis too. There also the properties of a function are in many ways determined by the nature of the domain. But in the first place this dependence is expressed less noticeably, or rather more conventionally; and in the second place the properties of the domains used in analysis are mainly elementary geometric properties known from another subject.

functions. It is also not coincidental that in 1899, at the very dawn of the emergence of the modern era in the theory of functions one of its founders, René Baire, wrote, "...in the order of ideas we have followed, all problems involving functions reduce to questions that belong to set theory; and in proportion to the degree of advancement or possible advancement of the latter, it becomes possible to solve a given problem [of the theory of functions—F. M.] more or less completely" [5, p. 121].

But Baire was only giving conscious expression to the dependence of the theory of functions on set theory. This dependence had been understood much earlier, and the idea of this dependence can be traced in the works of Hankel (1870), Smith (1875), Dini (1878), and many other mathematicians of the second half of the nineteenth century.

Up to now when speaking of sets we have meant point sets of ordinary Euclidean space, whose elements are geometrically identical points. But in the theory of functions more complicated manifolds are also considered—sets consisting of the same points of a Euclidean space, but "loaded" with some substance described by a distribution function. The substance is most often called mass, but this "mass" is not only physical, but also abstract; it may be a charge, a random event, or the like. It is just such complicated sets that form the basis of the theory of the Stieltjes integral and of differentiating one function with respect to another. To be sure sets of such a type do not form an independent branch of knowledge, but in fact the investigations in the theories just named are frequently interpreted precisely in the language of such sets.

The next general difference between the theory of functions and analysis consists of a change in the relation between the computational and conceptual aspects of the phenomena being studied. In this respect there is no clearly marked boundary between the two subjects just named, but one can nevertheless assert in a general way that the computational aspect predominated in classical analysis while the conceptual aspect predominates in the theory of functions. In the latter computational algorithms are more and more displaced, mainly by set-theoretic reasoning, and formulas by theorems. Such a distinction is characteristic not only for the disciplines under consideration; in a well-known sense it expresses the difference between the mathematics of the seventeenth and eighteenth centuries and that of the nineteenth and twentieth centuries.

The next difference, an important one in our view, is the change in the relationship between the two main operations of analysis and the theory of functions—differentiation and integration. Classical analysis arose mainly on the basis of the idea that differentiation was the primary operation, and it developed on that basis. For the theory of functions, however, the independence of the two infinitesimal operations is more characteristic, and there is even a noticeable tendency to make the operation of integration primary.

One final difference that we would like to point out is the prevalence of investigations of the summability of series and sequences in the theory of functions, in contrast to the prevalence of investigations into convergence in classical analysis. Divergent series had long been used in analysis, but only since the end of the nineteenth century have they become a "legal" tool for studying the properties of functions, and the study of them occupies an ever larger place in mathematical endeavor.

One can, of course, exhibit other differences, but they seem less essential to us, and therefore we shall not dwell on them.

Such is our response to the question posed above. Analysis and the theory of functions differ both in subject matter and in method. At the same time there is nothing like a sharp boundary between them. The basic elements of either of these subjects are also the basic component parts of the other, but in the theory of functions they are richer in content, more distinct in form, and raised to a higher level of abstraction. Thus we have the right to say that the theory of functions is classical analysis in a new stage of development; on the one hand, these are separate subjects, but on the other hand they are the same developing subject in different stages of its development.

1.4 The theory of functions and functional analysis

The relationship between functional analysis and the theory of functions of a real variable is approximately the same as the relationship between the latter and classical analysis. Indeed, while the fundamental object of study in the theory of functions is the concept of a function, the fundamental objects in functional analysis—direct generalizations of the concept of function—are the concepts of functional and operator. Just as the problems of classification and the study of the different types of function occupy a large place in the former, so in the latter the study of different types of functionals and operators (linear, self-adjoint, analytic, and the like) are a large division of the theory. There is also a study of individual functionals and operators corresponding to the detailed study of the properties of individual functions that are of special interest for one reason or another.

Furthermore, if the study of the properties of functions in the theory of functions is preceded by a large portion of the theory of sets in Euclidean space, it is also the case that the study of the properties of functionals and operators is preceded by a preparatory study of the properties of sets in topological, metric, Banach, Hilbert, and other spaces. Here, however, the difference between these two mathematical disciplines becomes apparent. While

Euclidean space is not an object of study in the theory of functions, its properties being considered given, known, and studied in another mathematical discipline, namely geometry, in functional analysis the study of the corresponding spaces themselves remains a problem, although a significant portion of the work of studying them has been taken over by another mathematical discipline of recent vintage, one which generalizes geometry—topology. The necessity of such a preliminary study of the properties of the regions on which functionals and operators are defined has long been recognized. Thus in 1912 Hadamard [2, p. 17] remarked that the classical concepts related to functions could be stated without any particular difficulty because the geometric properties of a line, a plane, or space "were known to us and seemed obvious, even those which, we now know, lead to great difficulties." He goes on to say,"A functional continuum, i.e., the manifold obtained by continuous variation of a function in all possible ways, actually does not suggest any simple image to the mind. Geometric intuition reveals nothing a priori on this score. We are obliged to remove this ignorance, and we can do this only analytically, constructing the functional continuum for applications—a chapter in set theory" [2, p. 17–18].

Even greater difficulties arise when one passes from functionals defined on sets of functions to functionals and operators with more general arguments. Continuing the words of Hadamard just quoted, Fréchet said in 1928, "...intuitively we do not know the properties of a functional continuum, but we know at least that the elements of this continuum are functions. But in general analysis [as Fréchet, following Moore, called functional analysis—F. M.] we do not even know the elements of the continuum being studied" [8, p. 271]. For that reason there is an even greater need for a preparatory study of the properties of abstract sets.

The fundamental operations on functions, differentiation and integration, become the corresponding operations on functionals and operators in functional analysis, and approaches to methods of introducing them in the latter repeat to a large degree the methods of introducing such operations in the theory of functions.

The relationship between these two operations in functional analysis brings out even more distinctly the qualities of their relationship in the theory of functions: separate study of them and a tendency to define differentiation using the concept of the integral. It is clearly of some interest that the germ of the idea of defining differentiation through the integral is contained in the earliest papers of Volterra on functional analysis: in 1887 he introduced the concept of the variational derivative, relying on the integral expression for the variation of a functional.

There are parallels between the studies of series of functions and series of functionals and operators. To be sure, in functional analysis the study of series seems not to have achieved the range and depth that characterize it in the theory of functions, but it has been carried on since the very first appearance of functional analysis and continues at the present time.

However in functional analysis more attention is given to the development of the concept of limit than in the theory of functions. This is natural, since the circle of questions in which this concept is necessary is immeasurably larger and more varied in the former. It is therefore not coincidental that the study of limiting passages of various types was a concern of a large number of the creators of functional analysis, starting with Fréchet (1906). Here again functional analysis fuses with topology.

In speaking of the relationship between the theory of functions of a real variable and functional analysis it is interesting to note that the first floor of the edifice of the latter was constructed not according to the model of the former, but in analogy with the theory of functions of a complex variable. In the first works on functional analysis Volterra chose the Riemannian theory of functions of a complex variable as the model for the new subject while Pincherle chose the Weierstrassian theory; and one cannot agree with Pincherle, who asserted [2, pp. 2–3] that Volterra was pursuing an analogy with the theory of functions of a real variable. Volterra was merely starting from the general definition of a functional formulated by analogy with the definition of a function of a real variable. This is exactly what Riemann had done in constructing his theory of functions of a complex variable. Volterra's subsequent reasoning was developed within the framework of the analogy with the Riemannian theory. Such a choice of the direction of investigation was in many respects determined by the fact that the theory of functions of a real variable needed for a different choice did not yet exist. The research of Borel, Baire, and Lebesgue began only in the second half of the 1890's, whereas Volterra and Pincherle had begun to work in the area of functional analysis a decade earlier, during the period of dominance of the theory of functions of a complex variable.

Volterra, Pincherle, and some of their contemporaries in the 1880's and 1890's obtained very many valuable results in functional analysis. But interesting as the projects and results of the founders of the new subject were, they did not evoke the necessary volume of investigations. Only after the theory of functions of a real variable was constructed in basic outline was Fréchet able to begin almost entirely afresh to build the edifice of functional analysis, now consciously guided by analogies with the theory of functions of a real variable rather than with the theory of functions of a complex variable. And it is precisely in the direction of Fréchet, rather than that of Volterra

1.4 The theory of functions and functional analysis 21

and Pincherle, that, on the whole, functional analysis began to develop,[7] and the development of these two subjects proceeded and continues to proceed in close interaction, sometimes so close that it is occasionally difficult to decide to which of them a given result belongs. We shall give just two examples of this.

In 1915 Fréchet [5] introduced the concept of an integral that generalizes the Lebesgue-Stieltjes integral. He conceived this integral precisely as the integral of a functional defined on an abstract set. According to his plan and in actual content—it is in fact functionals that are integrated—the theory of such integration should have been included in the sphere of functional analysis. However, in the book of Saks [2] the Fréchet integral is presented only as an abstract Lebesgue integral, although in some respects it is far from the Lebesgue constructions, especially in relation to the concept of measure. Since then it has usually been treated within the framework of the theory of functions of a real variable.

As a second example we mention the general theory of summation of series. Investigations into the summation of divergent series were already being carried out in the framework of classical analysis, but their real blossoming occurred at the very end of the nineteenth century, together with the birth of the subject usually called the modern theory of functions of a real variable. In the development of the theory of summation two stages may be distinguished. The first of these is to some degree summarized in the book of Hardy [6], in which many particular methods of summation are expounded in detail. In conceptual content it belongs completely to the theory of functions. However the development of matrix methods of summation gradually led to the necessity of studying infinite matrices by the methods of functional analysis[8] and to the application of the results obtained in this study to the theory of summation. Thus arose a general theory of summation, in which the ideas of the theory of functions and functional analysis were interwoven so tightly that it is impossible to say where one leaves off and the other begins. As an example, the book of Cooke, *Infinite Matrices and Spaces of Sequences* [1], devoted to the general theory of summation, is difficult to classify as belonging to one or the other of these subjects.

Thus, just as the theory of functions can be regarded as a generalization of classical analysis, so functional analysis appears as a generalized theory

---

[7] One may remark, however, that in recent times a there is a noticeable tendency to bring into functional analysis the ideas that guided Volterra and Pincherle in the sense of relying on analogies with the theory of functions of a complex variable.

[8] The theory of infinite matrices, of course, was generated not only, and not even primarily, by matrix summation methods.

of functions of a real variable. In the end functional analysis turns out to be simply classical analysis, only of an immeasurably more abstract order. Many mathematicians regard it in precisely this way, as is well illustrated, for example, by the books of Dieudonné [1] and Shilov [1].

1.5 The theory of functions and other mathematical disciplines

We have already mentioned the dependence of the theory of functions on the theory of point sets in § 3. This dependence is not entirely one-sided. These two areas of knowledge were actually born and developed together and are so closely connected with each other that, as we have remarked, they are often identified. Let us give some examples.

The birth of the Riemann integral was also the beginning of the theory of point sets. Indeed the very nature of the definition of this integral as the limit of finite sums under the single hypothesis that this limit exists required the study of conditions for integrability, which consist essentially in evaluating the measure of the set of discontinuities; and it was precisely the study of such conditions that led Riemann himself and then Hankel (1870) and other mathematicians[9] to the early ideas about the measure of sets. These ideas subsequently grew into the theory of Peano-Jordan measure.

Moreover it was the question of the uniqueness of the expansion of a function in a trigonometric series that served as one of the points of departure for Cantor's (1872) approach to the construction of set theory.[10]

In both cases the study of a particular function-theoretic question led to the ideas of the theory of point sets, and the introduction of these ideas made it possible to solve problems of interest in the theory of functions. One can give as many examples as desired of this kind, and quite a few will be pointed out below.

Although there are hidden currents of topology in analysis and the theory of functions of a complex variable, it is only with the construction of the theory of functions of a real variable and set theory that point-set topology, which became one of the leading mathematical disciplines of our day, began to take shape. And this applies not only to the general theory of topological spaces (Bourbaki [3, pp. 177–178]), but also to a large number of specific topological results. They grew as a result of the spread of function-theoretic ideas and methods to more general situations of a topological nature. On this topic we shall limit ourselves to a single example.

---

[9] For more details on this cf. Medvedev [2, pp. 43–45, 97–102]; for a more complete description cf. Hawkins [1, pp. 17–42, 55–61, 86–96, 146–154].

[10] For details cf. Medvedev [2, pp. 84–88] and Paplauskas [1, pp. 215–227].

The concept of a limit is fundamental in topology. Besides the fact that it arose and developed in analysis and the theory of functions, it is recognized that the concept of a generalized limit or the limit of a filter, which plays such a fundamental role in topology, was born in the Shatunovskii theory of real numbers and in the Moore theory of integration. Shatunovskii introduced it in his lectures of 1906–1907, which were published in lithograph form at that time. These lectures were not printed until 1923 [1]. Moore made his first approach to a generalized limit in 1915 in connection with the theory of integration [1]. In a joint article with Smith [1] in 1922 this limit was introduced in general form. A similar concept was introduced independently in 1923 by Picone.[11]

Probability theory was an anomaly among the mathematical disciplines for a long time, and Hilbert (1900) still classified it among the physical sciences [*Hilbert's Problems*, p. 34], despite its brilliant development as a genuine mathematical discipline in the works of members of the Russian school of probability in the second half of the nineteenth century.[12] The situation began to change after Borel, starting in 1905, began to apply new ideas in probability theory. "The modern development of probability theory took place to a large extent under the influence of the general ideas of set theory and the theory of functions of a real variable. These are the ideas that enabled probability theory to recognize itself as a mathematical discipline, develop a clear set of concepts within itself, and free itself from purely intuitive ideas and conclusions" (Gnedenko [2, p. 117]). The main credit for this belongs to Kolmogorov. The axiom system he proposed gained almost universal recognition. "Its success is explained by a variety of circumstances, among which we mention only the following: it corresponded to the general spirit of mathematics of the time, and it linked probability theory closely with measure theory, thereby opening before it a rich arsenal of well developed methods of investigation and making it possible to encompass in a single simple scheme not only the classical chapters of probability theory, but also its recently arisen concepts and problems" [Op. cit., pp. 118–119].

The connections of the theory of functions with algebra are also multifaceted. We shall limit ourselves to a few remarks on the cycle of papers by Carathéodory that began with the article "Entwurf für eine Algebraisierung des Integralbegriffs" [2] and concluded with the book *Maß und Integral und ihre Algebraisierung* [4], about which we shall have more to say in one of the

---

[11] We have not been able to consult the book of Picone *Lezioni di analisi infinitesimale* (Catania, 1923).

[12] On the Russian school of probability theory in the second half of the nineteenth century cf. Maistrov [1, Chap. IV].

following essays. The point of departure for his work is the fact that the proofs of the existence of various types of abstract integral depend in the final analysis on the possibility of carrying out the fundamental set-theoretic operations on sets. Hence arose the idea of passing to the abstract algebraic point of view in the theory of measure and integration, starting from axiomatically defined operations. With such an approach measure and integration receive new illumination and further development. Carathéodory's ideas were subsequently developed by many mathematicians in relation not only to measure and integration but also to a variety of other concepts of the theory of functions. When this was done, it was not only the theory of functions that profited, but also algebra, especially the theory of Boolean algebras.

One could write ad libitum about the connection between the theory of functions and other areas of mathematics; this would require quite a large amount of space, even if restricted, as above, to one or two illustrative examples in each instance. For that reason we shall limit ourselves to merely mentioning some other mathematical disciplines whose connections with the theory of functions are especially close. Naturally the theory of functions of a complex variable must be among the first mentioned, followed by the calculus of variations, the theory of differential equations, both ordinary and partial, number theory, and geometry.[13] If we speak of the indirect, as well as the direct, influence of the theory of functions, i.e., through the medium of other mathematical disciplines (probability theory, functional analysis, and the like), we would have to name mechanics, theoretical physics, cybernetics, and in general nearly every other area of the mathematical sciences.

Finally, it may be noted that the philosophical and general mathematical disputes about the problems of the foundations of mathematics in the twentieth century grew largely out of the polemic on the principles of set theory and the theory of functions that took place at the beginning of the century.

Such profound and varied connections of the theory of functions of a real variable with mathematics as a whole and with mathematical science are not coincidental. In its day mathematical analysis was just as closely connected with the latter. The theory of functions, being a generalization of classical analysis, has not only preserved these connections, it has strengthened, enlarged, and enriched them.

---

[13] For information on these connections in the works of Soviet mathematicians see: Bari, Lyapunov, Men'shov, and Tolstov [1, pp. 257–258].

Chapter 2
The history of the concept of a function

2.1 Some textbook definitions of the concept of a function

There is at present no single generally accepted definition of the concept of a function. The views of different mathematicians on this concept and the methods of introducing it diverge. We shall illustrate this in the present section. Before doing so we remark that there is also no single way of denoting the concept under consideration. The word *function* is most often used for this purpose, but the word *mapping* is quite common, and the terms *operator, operation, transformation,* and several others are used in various senses. Sometimes other connotations are given to nearly all of these terms.

The definition of a function that seems to be most widely used is given in terms of two undefined concepts: set and correspondence. Thus we read in a work by P. S. Aleksandrov, "If in some manner an element y of a set Y is made to correspond to each element x of a set X, we say that there is a mapping of the set X into the set Y, or a function f whose argument ranges over the set X and whose value belongs to the set Y" [4, p. 6].[1] The same definition is given by Kleene [1, p. 36].

In another variant of this concept the undefined term *correspondence* is also replaced by the undefined term *relation*. Before stating this variant we give certain explanations that will also be needed later.

Given two sets X and Y, there exists a unique set consisting of the ordered pairs (x, y), where $x \in X$ and $y \in Y$. It is denoted by the symbol $X \times Y$ and called the *Cartesian product* (or simply the *product*) of X and Y. To a relation $R(x, y)$ between $x \in X$ and $y \in Y$ there corresponds a certain property of an element $z \in X \times Y$; the subset of the product $X \times Y$ consisting of the elements for which this property holds is the set of pairs (x, y) for which the relation $R(x, y)$ is true. This subset is called the *graph* of the relation R.

We now formulate the concept of a *mapping*.

> Let X and Y be two sets, $R(x, y)$ a relation between $x \in X$ and $y \in Y$. R is said to be *functional in y* if for *every* $x \in X$ there

---
[1] This same definition is given for the concept of an operation in the book of Kantorovich and Akilov [1, p. 99], where certain restrictions are imposed on the sets X and Y.

exists *one and only one* $y \in Y$ such that $R(x,y)$ is true. The graph of such a relation is called a *functional graph* in $X \times Y$. Such a subset F of $X \times Y$ is therefore characterized by the fact that, for each $x \in X$, there is one and only one $y \in Y$ such that $(x,y) \in F$; this element y is called the *value of F at x* and is denoted by the symbol $F(x)$. A functional graph in $X \times Y$ is also called a *mapping of X into Y* or a *function defined in X and assuming values in Y* (Dieudonné, [1, p. 5]).

This definition is more formal and at the same time one notices in it the transition to the following formulation.[2]

If one wishes to reduce to one the number of undefined concepts in the definition of a function, that concept is most often taken to be the concept of a set. As in the preceding case, a function is introduced in terms of a set and a relation, but the latter is given a preliminary definition in terms of the former: binary relations are defined as subsets of the Cartesian product $X \times Y$. Once again a relation $R \subset X \times Y$ is called a *function* from X into Y if for each $x \in X$ there exists one and only one element $y \in Y$ satisfying the relation $R(x,y)$. In other words a function is a single-valued binary relation defined on a pair of sets X and Y. Such a definition is adopted, for example in the book of Kuratowski and Mostowski [1, pp. 64–65, 69]. The concept of a mapping is defined similarly in the book of Mal'cev [1, pp. 11–13]. The fact that in the final analysis only sets are used in this definition is expressed concisely as follows: "A triple of sets $\langle F, X, Y \rangle$ is called a *function* if 1) $F \subseteq X \times Y$ and 2) there are no pairs in F having the same first component and different second components" (Shikhanovich [1, p. 202]). It appears that this type of definition is nearly as widely used as the first, and there is a tendency for the second definition to supplant the first.

Sometimes one prefers to take the concept of a relation from mathematical logic, rather than the concept of a set, as the only undefined concept and introduce the latter in terms of the former. In that case one uses the fact that a set is usually regarded as some system of things possessing a certain property (unary relation, predicate).

Suppose given certain objects (things) x and y and a relation R between them. Every object x having the relation R with some object y is called a *predecessor* of the given relation R; every object y satisfying the relation $R(x,y)$ is called a *successor* of the relation.

A relation R is called a *one-many* or *functional relation*, or simply a *function* if, to every thing y there corresponds at most one thing

---

[2] It is expounded even more formally, and at correspondingly greater length, in the book of Bourbaki [1, pp. 71, 76].

x such that $R(x, y)$; in other words, if the formulas

$$R(x, y) \text{ and } R(z, y)$$

always imply the formula

$$x = z.$$

The successors of the relation R, that is, those things y for which there actually are things x such that

$$R(x, y),$$

are the *argument values* and the predecessors are called *function values* or, simply, the *values of the function* R (Tarski [1, p. 98–99]).[3]

If we do not require the minimum number of undefined terms, we can find yet another definition of function based on considerations of set theory and mathematical logic. To do this we take as undefined the concept of a set X consisting of elements x. For elements we introduce the concept of a propositional function $\varphi(x)$ with the explanation that $\varphi(x)$ denotes a certain condition, and if x satisfies this condition then $\varphi(x)$ is true. The Cartesian product of two sets X and Y is then introduced in the usual way, after which a propositional function of two variables $\varphi(x, y)$, $x \in X$, $y \in Y$, is defined as a propositional function of the one variable $z = (x, y) \in (X \times Y)$. A relation is defined as a propositional function of two variables and a mapping or function as a relation (Kuratowski [2, pp. 2–3, 7, 11–12]).

We now give two more quotations.

> By a *function*—or, more explicitly, a *one-valued singulary* function—we understand an operation which, when applied to something as *argument*, yields a certain thing as the *value* of the function *for* that argument. It is not required that a function be applicable to every possible thing as argument but rather it lies in the nature of any given function to be applicable to certain things, and when applied to one of them as argument, to yield a certain value. The things to which the function is applicable constitute

---

[3] For the sake of consistency we have changed the notation for a relation: instead of the notation xRy used by Tarski we have written $R(x,y)$ in our quotation. It should be kept in mind that for Tarski the argument of the function is y and its value is x.

the *range of* the function (or the *range of arguments of* the function) and the values constitute the *range of values of* the function. The function itself consists in the yielding or determination of a value from each argument in the range of the function... Of course the words "operation," "yielding," "determination" as here used are near-synonyms of "function" and therefore our statement, if taken as a definition, would be open to the suspicion of circularity. Throughout this Introduction, however,[4] we are engaged in informal explanation rather than definition, and for this purpose elaboration by means of synonyms may be a useful procedure. Ultimately, it seems, we must take the notion of function as primitive or undefined, or else some related notion, such as that of a class... (Church [1, p. 15]).

If in the words of the concluding sentence the possibility of accepting the undefinability of the concept of function is once admitted, Bochner speaks emphatically about this: "... in reality the concept of function is undefinable, and any would-be definition of it is tautological" [1, p. 184].

It is possible to give other formulations containing certain new shades of meaning. But the ones already given seem to exhaust the basic types of different approaches. The authors quoted are not the inventors of the definitions mentioned above, as will be discussed below. The purpose of this section was to describe various approaches to the concept of a function in modern mathematics, so that the historical picture of the development of this concept can then be studied within the framework of these approaches. In doing this we chose the authors of the indicated definitions to some extent deliberately in order to show that the most general concept of a function is not the object of study of any one mathematical discipline. It is of fundamental importance in analysis, the theory of functions, functional analysis, set theory, topology, algebra, and mathematical logic. On such a large scale it cannot be the subject of one comparatively small essay, since these definitions themselves took shape in different areas of mathematics, in some cases having no mutual connection, and different paths led to them. In what follows we shall mostly concentrate on the function-theoretic aspect of the concept that interests us, with only occasional digressions into other mathematical disciplines, mostly in order to make certain details more precise.

---

[4] The reference is to the introduction to the book of Church [1].

2.2 The concept of a function in ancient times and in the middle ages

Until fairly recently historians of mathematics did not look very far into the past in studying the development of the concept of a function. Thus, for example M. Cantor [1, pp. v, 215–216, 256–257] began the history of this concept with Leibniz and John Bernoulli. Zeuthen [1, p. 354] referred its origin to a slightly earlier period (1669–1670) and connected it with the name of Barrow. Hankel [1, p. 45] began with the concept of variable quantity in the writings of Descartes.

Later historians begin with ever earlier stages in the development of mathematical thought, with occasional fundamental differences of opinion, however[5]. For the remainder of this section we shall mostly adhere to the ideas developed by Yushkevich in the article "On the development of the concept of a function" [4], supplementing them somewhat from other sources.

The extremely abstract approaches to the concept of a function that were given in the preceding section suggest that it is a concept of extremely ancient lineage. Indeed, in these approaches the character of the functional correspondence (as a relation, an operation, a law, or a rule) is not specified in any way at all; it can be given by a table, a verbal description, a graph, a kinematical rule, an analytic formula, and the like.

> In one of these senses or another the concept of a function is already implicit in the simple rules for measuring the more useful figures and in the tables of addition, multiplication, division, etc., used to simplify computations (which are sometimes tables of functions of two variables) formulated at the dawn of civilization. At every step in the so-called elementary mathematics one encounters correspondences between numbers or quantities. To state this trivial fact does not lead to anything really useful. All distinctions are wiped out when this is done. All the interest lies in the genesis of the idea and the gradual recognition of it, in the specific content that it acquired with the progress of science and philosophical thought, and finally in the different roles it played at various stages of this progress (Yushkevich [4, p. 126]).

The words of Yushkevich just given refer to pre-Greek mathematics. If we now turn to the mathematicians and natural philosophers of Ancient Greece, although we do not find in them the idea of functional dependence distinguished in explicit form as a comparatively independent object of study,

---

[5] In this connection cf. Yushkevich [4, pp. 123–124].

nevertheless one cannot help noticing the large stock of functional correspondences they studied. The methods of defining a function enumerated above, except for the last (by an analytic formula), were all to be seen there, were quite profoundly developed, and were studied in various aspects.

Investigations into the simplest laws of acoustics, statics, and geometrical optics expressed in the form of a connection between different physical quantities; the study of dependences given by verbal descriptions, between segments of diameters and conjugate chords in the theory of conic sections; exhibiting kinematic rules to describe the behavior of curves; the study of various classes of curved lines; the compiling of tables of trigonometric functions—all these things are easy to recognize in the works of ancient Greek scholars that have come down to us.

These diverse functional dependences were studied from various points of view. The rudiments of a classification can be noted, for example, in the division of problems into planar, solid, and linear, or in Apollonius' classification of curves according to the form of the equations that define them. Approaches to the infinitesimal study of functions are quite distinctly expressed—the rudiments of differential and integral methods go back at least to Archimedes[6] Interpolation was applied in studying the correspondences given by a table; Euclid's study of the question of the minimal data needed to determine a problem can also be considered a progenitor of the idea of a function. It goes without saying that it was particular curves (functions) that were introduced, studied, and applied—the spirals of Archimedes and Pappus, the quadratix of Hippias, the conchoid of Nicomedes, and the cissoid of Diocles. In other words functions were handled very much as they were to be handled at a later time, when the concept of a function was stated in explicit form.

The idea of functional dependence was developed not only in the works of the ancient Greeks. Expressed mainly in the form of tablets of various kinds, it was significantly developed in the Babylonian astronomical texts of the Seleucid era (the third to first centuries B. C. E.)[7] and in the writings of Ptolemy[8]; and as far as is known at present, the ultimate achievement in this direction is found in the works of al-Bīrūnī dating from first half of the eleventh century. Al-Bīrūnī, besides having an extraordinarily large stock of various dependences of quantities given in the form of tables and by verbal description, seems to have been the first to study certain general questions related not to

---

[6] For the latter see Yushkevich, ed., *The History of Mathematics from Ancient Times to the Beginning of the Nineteenth Century* [1, pp. 117–127], henceforth referred to as *The History of Mathematics*.

[7] For more details see Hoppe, [1, pp. 150–151].

[8] Cf. Schramm [1, pp. 150–151].

particular functions or separate small classes of them but to all the kinds of functions he studied. To some degree he characterizes their domains of definition, determines extreme points and the values of the functions at those points, and studies the rules for linear and quadratic interpolation[9]. His ideas, however, evidently had no real influence on his contemporaries and successors.

A further development of the idea of a functional dependence occurs in Medieval Europe. Yushkevich goes so far as to say that the concept of a function appears explicitly for the first time in the writings of the natural philosophers of the Oxford and Paris schools in the fourteenth century [4, p. 13]. In the work of Bradwardine, Swineshead, Huytesbury, Oresme, and other scholars of this period mathematics is regarded as a fundamental tool for studying the physical world; the idea of natural laws as laws of functional type begins to take shape; "there arises... in very rudimentary form a theory of the variation of quantities as functions of time and the graphical representation of them, whose sources must perhaps be sought in Babylonian astronomy" (Bourbaki, [3, p. 217]). The general—albeit restricted—concept of which an equivalent was the later concept of a function is introduced in explicit form; and when this is done, special terms are introduced to denote the independent and dependent variables; the rudiments of infinitesimal considerations reappear in the study of functional dependences; the simplest forms of mechanical motion are studied as examples of such dependence; the germs of set-theoretic ideas appear.

The ideas of the medieval scholars just discussed became widely known in many countries of Western Europe. However they were then developed in a form different from that in which they originally appeared, and their application did not go beyond a small number of artificially posed problems; and "if in the development of these basic concepts of mathematics and mechanics the generalizing and abstracting thought of the medieval scholars advanced beyond the thought of their ancient predecessors, nevertheless in the wealth of specific mathematical discoveries the achievements of Swineshead or Oresme cannot begin to compare with the brilliant results of Archimedes or Apollonius" (Yushkevich [4, pp. 133–134]).

The reasons for this are various: the abstract Scholastic way of posing problems, the undeveloped state of the necessary mathematical apparatus, especially algebraic language, and mainly the absence of pressing needs of social utility—all this led to a situation in which the prefigurations of the medieval natural philosophers, in particular in relation to the concept of a function, generally did not enter the mainstream of mathematical thought, at least in the form in which they were developed. Their influence on the subsequent

---

[9] For more details on functional dependence in al-Bīrūnī see Rozhanskaya [1–3].

development of mathematics as a whole and especially on the development of
the idea of functional dependence remains insufficiently studied, but the elements of such influence can be traced quite distinctly in the works of Napier,
Galileo, Cavalieri, Descartes, Barrow, Gregory, Newton, Leibniz, and other
thinkers of the seventeenth century.[10]

2.3 The seventeenth-century origins of the concept of a function

However widely the diverse kinds of functional dependences had been studied
in the preceding periods of the history of mathematics, however clear the
"explicitness" in the introduction of the concept of function in the Middle
Ages, it was only in the seventeenth century that the latter began to acquire
the status of an independent and important mathematical concept. For it to
acquire this quality an enormous amount of preparatory work was necessary
in the most diverse areas of scientific endeavor.

First of all the scientific revolution of the seventeenth and eighteenth
centuries,[11] which was brought about by the evolution of socio-economic relations, led to a new, mechanical picture of the world. The transition to this new
picture had been prepared much earlier and from many different directions—
by the technological revolution, by the heliocentric system of Copernicus and
the subsequent advance of astronomy, by the new dynamics of Galileo, by the
development of statics and optics, etc. In this mechanical picture of the world
the essential, one might even say definitive, event was the concept of a law as
a dependence between variable quantities. Of this kind of dependence—the
laws established in the seventeenth century—one can exhibit many examples:
the period of revolution of a planet about the sun and its distance from the
sun, the velocity or distance and the time of motion, the period of vibration
of a pendulum and its length, the velocity of a liquid and the area of a cross
section, the angle of incidence and the angle of reflection of a ray of light,
the focal distance and the radius of curvature, and the like.[12] Each such dependence was studied by many people, by various methods, and in different
aspects. In order to do this, along with experimental apparatus and general theoretical considerations, mathematical methods were also developed:

---

[10] For more details on all the questions just touched on concerning the development of the concept of a function in the medieval period, cf. Yushkevich [4, pp. 130–134] and *History of Mathematics* [1, p. 270–283].

[11] Cf. Yushkevich [6] and *History of Mathematics* [2, pp. 7–21].

[12] "A complete enumeration of such types of laws would occupy many pages, even within the confines of the seventeenth century" (Yushkevich [6, p. 22]).

2.3 The 17th-century origins of the function concept

computational and descriptive devices were developed, analytic language was sharpened, and the necessary conceptual apparatus was established.

Since the concept of a law as a dependence between variable quantities became one of the central questions of the development of science, and mathematics was regarded as the most important, nearly universal method of knowing reality,[13] and since the scholars of the time were simultaneously physicists and mechanicists, astronomers and engineers, philosophers and mathematicians,[14] the concept of regularity was bound to be reflected in science. It grew into the concept of functional dependence.

What has just been said does not mean that the state of affairs consisted in the discovery of regularities in natural science one at a time, followed later by some rather general concept of a law of nature, and only afterwards, as a reflection of such a general notion, the introduction into mathematics of the concept of functional dependence. This would be a crude oversimplification of the actual state of affairs. We have already said that functional dependences had been studied almost since the very origin of mathematics, and it was only because the study of them had advanced rather far by the beginning of the seventeenth century—whether in the form of a verbal description or the graph of a curve, kinematically or in the form of tables—that it became possible to state the regularities in the natural sciences mentioned above. Thus the actual relationship between mathematics and the natural sciences was no mere dependence of the former on the latter, but a tight interweaving of each with the other, a mutual dependence in whose particulars one can hardly introduce a partial ordering, much less a linear one.

Nevertheless if the science and mathematics of the seventeenth century are regarded as a whole, one can assert that the general notion of a scientific law, a natural regularity, in natural philosophy preceded the explicit general concept of functional dependence, and that in the final analysis the latter was a reflection of the former.

To what extent this reflection was recognized by the mathematicians of the seventeenth century we find it difficult to say—this question, it seems, has not yet been studied by historians of science, but mathematicians at least have sensed the dependence of the concept of a function on the concept of a law. This is attested to by the fact that in the writings of Descartes, for example, we find both a general concept of natural regularity and a rather general

---

[13] Compare the corresponding declarations of Descartes and Galileo in *History of Mathematics* [2, p. 10].

[14] Ibid., pp. 11–13.

concept of a function[15]; the same can be said of Newton[16] and, apparently, many other scholars of the time.

But to sense the necessity of distinguishing the idea of functional dependence in mathematics as an object of independent investigation, as a reflection of the conception of regularity of natural phenomena—this is only half of the matter. Sufficient conditions for making such a distinction had to form within mathematics itself: the necessary modes of expression had to be formulated to describe these regularities, as well as the apparatus for studying them. And again it was in the seventeenth century that all this preparation was carried out: analytic geometry was invented, algebra and especially algebraic symbolism was developed, the device of power series began to be widely used, differentiation and integration were studied deeply. We shall not dwell on the this, referring the reader instead to the chapters of the second volume of *History of Mathematics*, while we take up a different question.

The methods of defining and studying functional dependences mentioned above continued to exist and develop. New tabular dependences were put forward—numerous astronomical tables, tables of logarithms, and the like. From Galileo to Torricelli and Roberval the kinematic method of defining and studying these dependences continued to develop; the concept of a curve continued to play the role of a universal method of expressing dependences. Nevertheless in the seventeenth century the representation of a function in the form of an analytic expression, a formula, gradually becomes the primary method of defining a function.

We have now become accustomed to consider the Bernoulli-Euler definition of a function—an analytic expression formed in some manner from variables and constants combined into a whole using the laws of mathematical operations—to be a narrow definition. However, if the Euler definition—"A function of a variable is an analytic expression formed in some manner from the variable and numbers or constants" [1, p. 4]—were interpreted literally, even the punctilious analyst of today would subscribe to it.

The point is that the words *analytic expression* that occur in this definition have almost the same indeterminate meaning as the word *function*. These words have changed their meaning as mathematics has developed; and if, for example, we take account of the results of Men'shov that any measurable function that is finite almost everywhere is the sum of a trigonometric series that converges to it almost everywhere, *a fortiori* if we take account of the fact that for any measurable function that is finite almost everywhere or

---

[15] For the concept of natural regularity in the writings of Descartes cf. *History of Mathematics* [2, p. 9]; the concept of a function in his writings will be discussed in the next section.

[16] For this cf. Section 4 below.

assumes the values $+\infty$ or $-\infty$ on a set of positive measure, there exists a trigonometric series converging to this function in measure, then it turns out that practically all the functions with which analysts deal today fall under the literal sense of the Euler definition given above.

But even if we agree with the scholars of the seventeenth and eighteenth centuries that analytic expressions include only analytic functions[17] the generality of Euler's definition is such that it actually encompassed all of the functional dependences known at the time, and Lagrange in 1813 (in the second edition of his *Théorie des fonctions analytiques*) asserted that every function can be represented by a power series [2, p. 16].

It was evidently this generality that predestined the analytic method of defining a function to become the dominant one. But, of course, it was not only the generality. An important role was also played by the loss in visualizability in studying rather complicated curves and the impossibility of discerning the properties of functions from a tabular representation of them containing large numbers of data, as well as the cumbersomeness of a verbal description that also did not permit mathematical operations to be performed on the function. It was precisely the analytic formula—a finite algebraic operation or a power series—that gave a visual form to the new object, the function, and made it possible to operate with it according to the laws of the algebra of the time and carry out infinitesimal operations.

Thus the extraordinarily large capacity of the analytic formula, along with its extraordinary flexibility in relation to analytic operations and its visualizability, which sometimes made it possible to discern the properties of a function directly from its definition as a formula—all this made it possible to distinguish an independent concept of a function in the form of an analytic formula represented by a finite algebraic expression or a power series.

2.4 Some particular approaches to the concept of a function in the seventeenth century

The concept of a function was established as the central concept of analysis by Euler only in 1748. But this event was preceded by long attempts at various approaches to this concept, choices of the most suitable method of defining it and the most reasonable notation and symbolism, the discovery of connections with the rest of mathematics and mathematical science, and the study of particular functions and classes of functions.

---

[17] For this cf. Markushevich [1, pp. 16–17]. But this was only the predominant tendency; there were exceptions, cf., for example, Yushkevich [4, pp. 146–147].

36 Chapter 2 The history of the concept of a function

The discovery of the analytic method of defining a function is now connected with the names of Fermat and Descartes (Yushkevich [4, pp. 136–137], Natucci [1, pp. 89–90]), especially the latter, who was the first to point out clearly that an equation containing two variables makes it possible to determine all the values of one of the variables from given values of the other (Descartes [1, p. 34]). Two points in their conception of a function should be noted.

First of all, a geometric representation—a curve—is still in the foreground of their thought, and the analytic formula is only a way of describing it, convenient and useful, but nevertheless auxiliary. Thus Descartes speaks not of classifying analytic expressions, but of classifying curves [1, p. 40–44]. Such a tendency remained in mathematics for a long time.

Second, a function, as a rule, is thought of as given implicitly, in the form of an equation between two variables, and not in the form $y = f(x)$, i.e., a formula solved with respect to one of the variables. This also is a tradition that was preserved for quite a long time. It is evidently explained by two factors. The first follows from the fact that the concept of a function was shaped by the concept of regularity in natural science. Indeed a large part of the latter took the form of an implicit equation between variables. Even in the cases where we now see an explicit expression of one quantity in terms of another, the mathematicians and mechanicists of the seventeenth century, following their own tradition, which goes back at least to Eudoxus, spoke of a ratio of quantities: they did not define area or volume, velocity or pressure, but rather the ratio of areas, volumes, etc. The second factor was evoked by the fact that the representation of a functional correspondence as a formula had its own origin in algebra, and algebra at the time was thought of as the study of equations.

The bridge from the primacy of the curve to the primacy of the formula and from the implicit to the explicit definition of a function is apparently the creation of Newton.

The concept of a function in Newton has been studied in many works and in many aspects.[18] However, it does not seem to have been studied from this point of view, and therefore we shall permit ourselves to go into more detail.

In a work written as early as 1665–1666 Newton [1], though still beginning with the geometric representation of a function as the variable ordinate of a curve depending on an abscissa, relies basically on the analytic representation of the function. He denotes the abscissa, as we do today, by the letter x and the ordinate by y. The expression of y in terms of x he writes in

---

[18] Cf., for example, Yushkevich [4, pp. 139–141].

2.4 Some particular 17th-century approaches

the form $y = f(x)$, not, of course, using the symbol f, but simply giving it a specific analytic expression in various special cases, for example,

$$y = \frac{1}{1+x^2}, \quad y = x^3 + \frac{\sqrt{x}-\sqrt{1-x^2}}{\sqrt[3]{ax^2+x^3}} - \frac{\sqrt[5]{x^3+2x^5+x^{2/3}}}{\sqrt[3]{x+x^2}-\sqrt{2x-x^{2/3}}},$$

$$y = \frac{a^2}{b} - \frac{a^2 x}{b^2} + \frac{a^2 x^2}{b^4} - \cdots$$

It is these analytic objects that are the main subject of his investigation. Geometric considerations are either not mentioned or introduced only in the most necessary cases. Thus for example in calculating the integral $\int \frac{\sqrt{1+ax^2}}{\sqrt{1-bx^2}}\, dx$ Newton remarks only in passing (in parentheses on p. 8) that the quadrature of the curve $y = \sqrt{1+ax^2}/\sqrt{1-bx^2}$ gives the length of an arc of an ellipse. Otherwise he treats this function as a purely analytic expression unconnected with any geometric image.

Only in the section "Application of what has been expounded to other problems of the same type" does Newton explicitly appeal to geometric reasoning. For example, in finding the differential of length of an arc of a circle, he resorts to a drawing, and, once having obtained from it an analytic expression for the differential, proceeds from then on as before, purely analytically. But, having found the series for arcsin x (not, of course, in this notation), he obtains the series for the sine without the use of geometry—by inverting the series

$$x + \frac{1}{2}\frac{x^3}{3} + \frac{1 \cdot 3}{2 \cdot 4}\frac{x^5}{5} + \frac{1 \cdot 3 \cdot 5}{2 \cdot 4 \cdot 6}\frac{x^7}{7} + \cdots$$

His approach to $\cos x$ is even more complicated and again uses no geometric images: he finds the series for $\cos x$ from the formula $\cos x = \sqrt{1 - \sin^2 x}$ and the series for the sine. He proceeds similarly in other cases also.

It was said above that in his concept of a function Newton started from a geometric representation—some given curve. But the central point of his paper [1] is precisely the replacement of the initial image with a different one, the analytic. Indeed, after discussing the first two rules relating to the quadrature of simple curves (i.e., to the integrals $\int ax^{m/n}\, dx$ and $\int \sum a_i x^{m_i/n_i}\, dx$), he goes on to speak of the quadrature of "all other curves," putting those words into the title of the main section of his work; and for that purpose he thinks of every function/curve as given analytically, in the form of a (finite or infinite) power series, from which termwise integration gives the desired quadrature. Consequently his original image of a function—a geometric curve—is turned into an analytic image, a power series acting as a representative of any functional dependence. It was such an approach that enabled him to solve the problem of the quadrature of the curves that mathematicians of the time were interested in.

In the second of his papers on analysis Newton [2] deepens this approach, and the concept of a function as an analytic expression receives further development. Before discussing this we must make another remark.

Newton was also a physicist. Mathematics interested him not only for its own sake, but also as a tool for studying the regularities in mathematical science. The latter, as we have said, were most often expressed not in the form of an explicitly given dependence of one quantity on one or more other quantities—in mathematical language, not in the form of an explicit function—but in the form of an equation or a system of equations. Therefore the study of such dependence is of great interest for applications, and Newton frequently engages in it. This is particularly evident in [2], although it is noticeable also in [1].

In his *Fluxions* we encounter the kinematization of the concept of function spoken of by Yushkevich [4, p. 140]. But this kinematization is again rather an outer garment. The inner content with which Newton invests this concept, with which he primarily works, is some analytic expression or other.

Here we encounter a still larger class of functions defined analytically, and the stock of analytic methods enlarges as well. If in the first paper functions were defined in explicit form or in the form of an implicit algebraic equation, in the second we meet mostly the study of functional dependence defined in the form of differential equations (although the stock of functional dependences described by the first two methods is significantly larger). Here Newton, evidently in accordance with the method of describing scientific regularities, prefers to study functional dependence in implicit form. Even in those cases when the expression of this dependence in the form $y = f(x)$ causes no difficulty he preserves the implicit expression for it.

What has just been said does not mean that Newton is not interested here in functions that are analytically defined in explicit form. On the contrary, besides the significant enlargement of the stock of them in the earlier form, there is here an obviously new quality in the very character of analytic expression. To be specific, if in the first paper Newton mostly considered expressions in which the exponents were integers (except for the function $y = ax^{m/n}$), in the second work the numbers, as a rule, are arbitrary parameters. As a consequence of this there is a loss of the geometric visualizability that is usual for representing a function in the form of a specific curve whose properties can be seen by relying on geometric intuition. Indeed, what visual geometric image can be connected, for example, with the "curve"

$$y = \frac{2\theta e z^{\theta-1} + (2\theta - \eta)fz^{\theta+\eta-1} + (2\theta - 2\eta)gz^{\theta+2\eta-1}}{(e + fz^\eta + gz^{2\eta})2\sqrt{e + fz^\eta + gz^{2\eta}}},$$

where the letters η and θ "annex'd to the quantity z denote the number of dimensions of the same z, whether it be integer or fractional, affirmative or negative" (p. 99). But then this is only one of many examples contained in this work of Newton's. Such a function can be called a curve only conditionally, in a more abstract sense than that which is usually connected with the traditional curves such as the ellipse, the cycloid, and the like. This is a family of specific curves, moreover an infinite one.

The idea that it was necessary to give up synthetic reasoning in the study of functions matured in Newton no later than 1676. By that time Newton had begun to use the special term *ordinate* to denote the concept that corresponds to our notion of an explicit function given by an analytic expression. Where earlier he seems not to have used this word in this sense, in a letter to Leibniz of 24 October 1676 his term *ordinate* is best translated by the word *function*. He writes as follows: "Let the ordinate be" [3, pp. 116, 135], "Let the ordinate be $\frac{a^5}{z^5}\sqrt{bz+zz}$" (pp. 116, 135), and the like. And to use Newton's expression, there are "an infinite number" (pp. 119, 138) of such "ordinates," including those of the type mentioned above, where the expression for the function contains a variable in the exponent that is a parameter. Analytic images of this type become impossible to study by the old synthetic methods. Therefore, in calculating the integrals of such "ordinates" Newton writes candidly to Leibniz: "I would not, of course, have been able to obtain any of these results until I renounced the contemplation of figures and reduced the whole matter to the study of ordinates alone" (pp. 120, 138). A significant confession! To find the integral of even moderately complicated functions it is necessary to give up the study of geometric figures and study functions given in the form of analytic expressions.

Finally, Newton studies even more general *ordinates* in [4], for example,

$$y = x^{m-1}[f(x^n)]^{p-1}[g(x^n)]^{q-1}h(x^n),$$

where

$$f(x^n) = f_0 + f_1 x^n + f_2 x^{2n} + f_3 x^{3n} + \cdots,$$
$$g(x^n) = g_0 + g_1 x^n + g_2 x^{2n} + g_3 x^{3n} + \cdots,$$
$$h(x^n) = h_0 + h_1 x^n + h_2 x^{2n} + h_3 x^{3n} + \cdots,$$

and f_i, g_i, and h_i are constants (p. 71); and of course it is necessary to work with them without the support of geometry, and *a fortiori* not as kinematic images, but according to the rules of analytic transformations, which he does successfully, in particular compiling extensive tables of derivatives and primitives, defined analytically.

Thus one can say the following about Newton's concept of a function. Although he attempted to describe this concept in the language of geometry and mechanics, he actually worked with functions as analytic expressions composed of variables and constants. To exhibit such expressions he used the terms *curve* or *ordinate*, but by no means preserved the notion of a specific curve or the ordinate of a specific curve. He seems to have given no special definition of *curve* or *ordinate*, but from the stock of "curves" or "ordinates" at Newton's disposal, which was collossal by the standards of the time, it was comparatively simple to obtain the definition of a function as an analytic expression, as was done by John Bernoulli in 1718, understandably relying for this not only on Newton's papers (which, as it happens, were not all known to him) but also on the investigations of Leibniz, James Bernoulli, and others, as well as his own.

What has been said about Newton's concept of a function does not, of course, mean that he was the only mathematician in whose writings one can trace the transition to the concept of a function that came about with John Bernoulli and then was adopted and deepened by Euler. The elements of this transition are apparent in the writings of Mengoli, Wallis, Barrow, Gregory, and many others. During this same period the properties of specific functions (for example, the logarithm) and classes of functions (algebraic, as well as trigonometric and other transcendental functions) were being studied, and their connections with one another were being established, etc. To write about this in detail would mean to write a general history of mathematical analysis in the seventeenth century. Therefore we shall limit ourselves to what has already been said and add only a few words about the term *function* itself.

It is usually said to date to the writings of Leibniz (in his manuscripts of the year 1673 and in his publications of 1692 and 1694) (Yushkevich [4, pp. 141–142]). Some historians, as we have said at the beginning of Section 2 of the present chapter, attempt to date the history of the concept of function itself from this year. It should be noted that the content (expressed rather unclearly, to be sure) with which Leibniz invested this term was farther from the definition of John Bernoulli than Newton's *ordinate*, and it is John Bernoulli who deserves the credit for giving this term its modern meaning.[19]

2.5 The Eulerian period in the development of the concept of a function

We shall provisionally use the term *Eulerian period* to denote the period from the appearance of the definition of a function in the writings of Bernoulli in 1718 until the publication of Fourier's *Théorie analytique de chaleur* (1822).

---

[19] For more details cf. Yushkevich [4, pp. 142–143].

2.5 The Eulerian period

Following a lively discussion of the concept of a function in the Leibniz–John Bernoulli correspondence[20] Bernoulli's definition finally appears in print: "A function of a variable is defined here as a quantity formed in any manner from this variable and constants."[21] In this same place we find a nearly modern symbolism for denoting a function: φx.

If this definition is read literally, one can find in it the content that is now invested in the words "Dirichlet concept of a function." Indeed it speaks of a "quantity" formed in any manner from a variable and constants and says nothing about any obligation to define the quantity as a formula. However this last meaning followed from the content of Bernoulli's work and it was in that form that mathematicians adopted it. It was evidently for that reason that Euler, in nearly recreating the Bernoulli definition thirty years later, replaced the word *quantity* with the words *analytic expression* [1, p. 4]; considerably later Lagrange proceeds exactly the same way [2, p. 51]. Bernoulli's indeterminate "quantity," which, taken together with the arbitrary manner of defining it, made it possible to treat the concept of a function more generally, too indeterminately for that time, was thereby amended.[22]

Both the Bernoulli definition itself and even more its "amendment" by Euler and Lagrange had been preceded by the introduction of the universal apparatus of the time for representing a function analytically—the Taylor series (1715), which seems to have been in the possession of Gregory even earlier (1672) (*History of Mathematics*, [2, p. 166]) and was actually known, though in a slightly different form, to Leibniz and John Bernoulli (Ibid., pp. 273–274). Thus the introduction of the Bernoulli-Euler definition was essentially a recognition of the fact that mathematicians had at their disposal a method of representing functional dependences analytically that enabled them to describe all the functions known at the time. It was this recognition that constituted Bernoulli's definition.

This definition did not gain immediate general acceptance. Apparently until the publication of Euler's book [1] it was customary to think of functional dependence primarily either in a purely geometric way, as some line, surface, or solid, or as some mixed geometric-kinematic structure, when lines, surfaces, and solids were thought of as generated by the motion of points, lines, and surfaces. As evidence of what has just been said we refer to the books of Reyneaux [1, p. 151], Agnesi [1, p. 2], and others. But in the actual study of functions, especially questions of differentiation and integration,

---

[20] Ibid., p. 142.

[21] Quoted by Yushkevich [4, p. 143].

[22] Naturally only slightly. As Yushkevich properly remarks [4, pp. 145–146], the range of the term "analytic expression" also remained rather indefinite.

mathematicians always tried to express a functional dependence as some formula composed of variables and constants. Thus, for example, De Moivre [1], having functions in mind, speaks of a quantity or ordinate, but always writes them in the form of some analytic expression, and it is in that form that he studies them. Even Maclaurin [1], who clung to the geometric-kinematic representation of functional dependence longer than anyone else and perhaps obtained more important results than any other of his contemporaries with that approach, was obliged to bring in the Bernoulli definition of a function (p. 578) and use it in the study of functions. It was precisely in this way, algebraically as he put it, that he obtained his series—a somewhat different universal analytic method of representing functional dependence.

The opinion was sometimes expressed that it was Euler who made the concept of a function central in analysis. Thus Hankel wrote: "It was Euler who first established it and made it the foundation of all of analysis, and a new era in mathematics was thereby begun" [1, p. 15].[23] This is not true if we interpret the opinion just quoted too literally. In reality the concept of a function lay at the foundation of analysis long before Euler; it merely bore a different name—line, surface, or solid—and was studied by different methods. But if we understand a function to mean an analytic expression, not necessarily connected with any geometric considerations, this opinion describes the state of affairs exactly. Having precisely John Bernoulli's definition of a function in mind (with his own amendments to it), Euler expressed the fundamental idea that "all of infinitesimal analysis revolves about variable quantities and their functions" [1, p. VIII] and he adopted the Bernoulli definition as the point of departure.

But since the term *variable magnitude* or *variable quantity* occurs in the Bernoulli definition, Euler preceded the definition of the concept of a function with an explanation of this term. The Euler explanation deserves attention for the following reasons.

When one has in mind a functional dependence expressed in the form of certain geometric relations, some geometric object is always behind it, represented with a greater or lesser degree of distinctness, and defining what can be provisionally called the domain of definition and the range of values of the function. There is no pressing need to exhibit them specially; they can be "discerned," so to speak, using the geometric image itself. Such is not the case when, as with Euler, "a function of a variable quantity is an analytic expression formed in any manner from this variable quantity and numbers or constant quantities" [1, p. 4]. No specific geometric or kinematic image lies behind the proposed analytic expression to enable us to judge what values

---

[23] Cf. also Bashmakova and Yushkevich [1, p. 477].

to give to the letters occurring in the expression (we recall the Newtonian analytic expressions with parameters that did not permit the usual images of the time to be associated with them). These values had to be described somehow.

The descriptions given by Euler are interesting in two respects.

First, "a variable quantity is an indefinite or universal quantity containing within itself all possible values" [1, p. 4]. In other words for Euler a variable quantity is not a flowing, changing quantity, as it was, for example, for Maclaurin [1, pp. 52–57, and many other places], but rather some real structure: "just as the concepts of species and genera are formed from the concepts of individuals, so a variable quantity is a genus in which all the definite quantities are contained" [1, p. 4].

Second, he writes "a variable quantity encompasses the collection of all numbers... absolutely all numbers can be substituted for it." Consequently Euler regards functions of one variable[24] as being defined, as we would nowadays say, on the entire real line. He writes, "... the meaning of a variable quantity will not be exhausted until all definite values are substituted for it." [1, p. 4]. Moreover, "not even zero and imaginary numbers are excluded from the values of a variable quantity" (Ibid). Hence for Euler the domain of definition of a function is in general the entire plane.

He treats the range of values of a function exactly the same way, for "a function of a variable quantity is itself a variable quantity" [1, p. 4].

Thus for Euler a function is in general an analytic function defined on the entire complex plane and assuming any complex value. There is no such open and direct inference in [1] or in any other work of Euler; however, such an inference can be made both from the totality of statements given and from the general content of many of his works.

A great deal could be said about the narrowness and at the same time the extraordinary breadth of Euler's concept of a function (especially taking into account the fact that this book of Euler's appeared in 1748) and about Euler's contribution to the development of this concept. We remark only that in [1] Euler restricted himself to real-valued functions and usually functions of a real variable; but in other works he also made a broad study of functions of a complex variable with complex values[25]; in [1] he not only stated the concept of a function as an analytic expression, he also developed it profoundly in many diverse directions, summarizing what had been done by his predecessors and adding much of his own.[26]

---

[24] And it is the study of these objects that forms the subject of his book [1].
[25] For this cf. Markushevich [1, p. 15–48].
[26] Besides the references in the preceding footnote, see also Timchenko [1, pp. 260–314].

2.6 Euler's contemporaries and heirs

During the seventeenth and eighteenth centuries, when mathematical analysis was growing into a large independent scientific discipline the question of providing a foundation for it was almost never off the agenda. Various methods were invented for this, whose supporters engaged in very heated discussions; in the course of these discussions the fundamental concepts of analysis were made more precise. We do not intend to dwell on this[27] and we mention these disputes only because one of these approaches is inseparably linked with the concept of a function discussed above.

Since the concept of a function had become the central concept of analysis by the middle of the eighteenth century and the definition of a function in the form of an analytic expression represented by a power series had also become generally accepted, there was a natural attempt to take this definition itself as the starting point for the foundation of all analysis. This is the way in which Lagrange proceeded.

The idea of such an approach was first stated by him in the paper "Sur une nouvelle espèce de calcul relatif à la différentiation et à l'intégration des quantités variables" [1], published in 1772. In this paper Lagrange proposed a new purely formal proof of Taylor's theorem on the expansion of a function in a power series. Noting that in the case of an algebraic function this expansion has the form

$$f(x) = f(a) + \frac{f'(a)}{1!}(x-a) + \frac{f''(a)}{2!}(x-a)^2 + \cdots + \frac{f^{(n)}(a)}{n!}(x-a)^n,$$

where all the successive functions $f'(x), f''(x), \ldots$ are obtained from the given function by the same law and are its successive derivatives, he concluded that in this case there is no need to appeal to any principles except those of algebra alone to find the derivatives. But since integration was only the inverse of differentiation for him, ordinary elementary algebra also sufficed for integrating functions (for the time being only algebraic functions). Thus one could provide a foundation for the infinitesimal operations on algebraic functions without resorting to the concepts of infinitely small, limit, indivisibles, or similar controversial ideas.

But the basic object of analysis—the function—was representable by a power series, the Taylor series, and therefore the entire theory of functions— the analysis of the time with all its applications—could be based on the same

---

[27] There is a voluminous literature in the history of science devoted to the problems of the foundations of analysis in the seventeenth and eighteenth centuries; we note, for example, the book of Carnot [1] and the paper of Yushkevich [1]; of the more recent studies we note only the article of Kitcher [1].

considerations. His papers [2] and [3] are a development of precisely these considerations.

We have no need to go into the details of Lagrange's approach here.[28] We merely note that Lagrange's enterprise was grandiose, not only in conception, but also in execution. It is easy for us now to criticize him for his restricted concept of a function, lack of rigor in argument, and the like; but Lagrange was able to fit almost all the known functional dependences of the time into this "restricted" concept and study them from a unified point of view, adding much that was new to the analysis of infinitesimals.

Nevertheless the time was really past when a function could be thought of as an analytic expression defined by a power series, especially a function of a real variable. From the middle of the eighteenth century on it gradually became clearer and clearer that such a notion of a functional dependence was too restricted and that it must be broadened. Two points should be noted in connection with this broadening.

The first point is that the needs of astronomy and mathematical physics had led to a new analytical method for depicting functional dependences—trigonometric series. "The accumulation of partial expansions [of functions in trigonometric series—F. M.] in the eighteenth century was evoked mainly by various problems of applied mathematics, and up to the time of Poisson advances in the field of trigonometric series depended almost entirely on problems of an applied character" (Paplauskas, [1, p. 7].) The newly discovered mechanical regularities (in the motion of heavenly bodies, in the theory of vibrations, in the theory of heat, and the like) could no longer be expressed in such a simple form. They could not be written in the form of a finite or infinite power series. It became necessary to apply new methods for the analytic description of these regularities, and this was done on an ever increasing scale. A significant role in all this was played by the famous polemic on the vibrating string.[29]

The second point that we would like to emphasize is that mathematicians returned, sometimes in spite of their own convictions, to the old geometric representation of a functional dependence in the form of a curved line or surface, moreover a line traced by a moving point or a surface traced by a moving line. From the middle of the eighteenth century on, when Euler, d'Alembert, D. Bernoulli, Lagrange, Monge, and many others have a function in mind, they once again speak of a curve or surface, not only of a formula, indeed not even primarily of a formula. To the majority of them a formula begins

---

[28] They are discussed by Dickstein [1], Timchenko [1, p. 314–350], and Markushevich [1, p. 41–45].

[29] For this cf. Paplauskas [1, pp. 17–38]; for more details cf. Timchenko [1, pp. 473–515].

Chapter 2 The history of the concept of a function

to seem too restrictive to convey every imaginable geometric/kinematic functional dependence in the one-dimensional case—a curve that can be described by a free movement of the hand. Only D. Bernoulli dares to assert that every such curve can be represented in the form of some analytic expression, formed no longer by a power series, however, but by a trigonometric series. It is characteristic in this situation that he starts from physico-mechanical considerations, relying on what seemed to him the reasonable proposition that the motions that exist in nature must reduce to sums of simple isochronal motions (Timchenko [1, pp. 493–496], Paplauskas [1, p. 26]). It is useful to quote a place in a letter from him to Euler. Having in mind the representation of an arbitrary curve by a trigonometric series, D. Bernoulli wrote:

> Nevertheless, here is my own opinion on this problem: we have proved that every curve expressible by the equation
>
> $$y = \alpha \sin \frac{\pi x}{a} + \beta \sin \frac{2\pi x}{a} + \gamma \sin \frac{3\pi x}{a} + \text{etc.},$$
>
> satisfies the conditions of the problem. But why could we not say that this equation encompasses all possible curves? Why could we not force the curve to pass through an arbitrary number of points whose location is preassigned by using the arbitrariness of the collection $\alpha, \beta, \gamma, \ldots$? Does an equation of this nature have less generality than the indeterminate equation
>
> $$y = \alpha x + \beta x x + \gamma x^3 + \text{etc.}?$$
>
> Wouldn't your beautiful proposition that every curve has the property studied in our problem be proved in this way? Thus to solve your problem, *Given any initial position, find the subsequent motion*, I say that one need only determine the collection $\alpha, \beta, \gamma, \ldots$ in such a way as to identify the given curve with the curve corresponding to our indeterminate equation. We will then immediately obtain the partial isochronous vibrations of which the desired motion is composed. Using this method I have been able to solve the problem of determining the motion of a taut string with an arbitrary number of loads of arbitrary mass situated at any distances from one another, and it seems to me that this problem is even more general than yours.[30]

A rather curious situation arose. On the one hand, starting in the second half of the seventeenth century and throughout the eighteenth century the ever

---

[30] Quoted by Timchenko [1, pp. 495–496].

2.6 Euler's contemporaries and heirs

increasing analytization of the concept of a function can be clearly traced (Newton, Leibniz, John Bernoulli, Euler, d'Alembert, Lagrange, and others); the geometric notion of a functional dependence retreats into the background, is even carefully avoided, for example by Euler and Lagrange.[31] On the other hand, when the disagreement over the vibrating string problem made it necessary to penetrate more deeply into the nature of the concept of a function, when functions of a more general nature were encountered, these same mathematicians, Euler and Lagrange, now basing their argument on geometric interpretations, came out against the essentially analytic approach of D. Bernoulli (which, to be sure, was not yet developed and was insufficiently grounded) even though they did so in contradiction to their own fundamental views.

Euler found the way out of this contradiction in formulating a new and extremely general concept of a function to which many circumstances led him.[32] In 1755 he expressed himself as follows: "When certain quantities depend on others in such a way that they undergo a change when the latter change, then the first are called *functions* of the second. This name has an extremely broad character; it encompasses all the ways in which one quantity can be determined using others" [2, p. VI].

In essence this is the definition that became known as Dirichlet's definition. Naturally one can argue about what Euler in 1755 understood by "all the ways in which one quantity can be determined by others," but the fact that at least some mathematicians adopted it in precisely this sense is attested by the example of Lacroix.

In his fundamental treatise [1] he says the following: "The older analysts in general understood a function of a quantity to be all the powers of that quantity.[33] Subsequently the meaning of this word was broadened and applied to the results of algebraic operations: thus any algebraic expression formed in some way from sums, products, powers, and roots of these quantities was also called a function. Finally the new ideas that arose as a result of the progress of analysis yielded the following definition of a function: *Any quantity depending on one or more other quantities is called a function of the latter, whether or not we know which operations must be performed in order to obtain the former*

---

[31] Such a retreat from geometric interpretations by Euler and Lagrange occurred not only in relation to the concept of a function, but also in many other questions, including mechanical questions.

[32] Cf. Yushkevich [4, pp. 147–148].

[33] Lagrange [2, p. 15] and d'Alembert made the same assertion (Natucci [1, p. 91]). Eneström—a rather scrupulous historian of mathematics—considered this assertion unfounded, although it may be that someone did indeed use the word *function* in this sense.

from the latter" [1, T. 1, p. 1].[34]

Lacroix not only introduced this definition into his basic course [1] but also stated it in some more elementary textbooks on mathematical analysis, as well as in the second edition of the treatise under discussion. It is essentially a reformulation of the definition of Euler given above, but with emphasis on the lack of constraint on the operations that must be carried out on the values of the argument in order to obtain the value of the function.

Lacroix did not refer directly to Euler in connection with the definition he gave. It is possible that he did not take it directly from Euler. The fact is that his friend and collaborator Condorcet was in possession of a concept of function that is in some respects even more general than that expressed in the Euler definition. In 1779, revising his 1765 course on integral calculus, he began as follows: "I assume that I have a certain number of quantities $x, y, z, \ldots, F$ and that for each definite value of x, y, z, etc., F has one or more definite values corresponding to them; I say that F is a *function* of $x, y, z, \ldots$".[35]

In saying this Condorcet consciously emphasizes that the method of defining a function may not necessarily require an explicit expression as an analytic formula or even in the form of an equation defining it implicitly. He says frankly that the dependence of F on $x, y, z, \ldots$ may hold "even when I know neither the way in which F is expressed [by an analytic formula—F. M.], nor the type of equation [defining it implicitly—F. M.]" (Granger [1, p. 85]).

The fact that this definition was not published at the time (Condorcet did not complete the revision of his course of integral calculus), does not mean that it left no trace in the history of mathematics. Lacroix could have known about it from conversations with his friend or even from the text of the unpublished manuscript.

Thus the notion of a function as an arbitrary dependence between variables, not necessarily defined by any analytic relation, was quite widespread at the turn of the nineteenth century. And even if the words "arbitrary dependence" had a narrower meaning for Lacroix than was given to them later, the very fact that his formulation was widespread made it possible to give it a broader content.

The words "arbitrary function" were given a very general meaning by Fourier by 1822 at the latest. In his epochal work *Théorie analytique de*

---

[34] It is curious to see that in his book [2] published in the same year Lagrange does not go beyond the definition of John Bernoulli, considering it to be the generally accepted one [2, p. 15].

[35] Quoted by Granger [1, footnote on p. 85]. For more details on Condorcet's definition cf. the article of Yushkevich [7].

chaleur [1] the notion of functional dependence that is widely believed to be connected with the name of Dirichlet, it seems to us, lies at the foundation of his entire investigation. One can exhibit many pages of his work on which he speaks more or less definitely about such a concept of functional dependence, for example, pages 10, 207, 209, 224, 232, etc; we shall give here only one excerpt from this to confirm what has been said:

> A general function $f(x)$ is a sequence of values or ordinates, each of which is arbitrary... It is by no means assumed that these ordinates are subject to any general law; they may follow one another in a completely arbitrary manner, and each of them is defined as if it were a unique quantity.
> From the nature of the problem itself and the analysis applied to it, it might seem that the transition from one ordinate to the next must proceed in a continuous manner. But then we are talking about special conditions, whereas the general equality (B)[36] considered by itself, is independent of these conditions. It is strictly applicable to discontinuous functions also [1, p. 500].[37]

It is not coincidental that Fourier gave such a general meaning to the term *function*. He was writing a work on mathematical physics. The phenomena he was studying were described in the language of partial differential equations, and the general integrals of such equations contain functions that assumed the most diverse forms in specific cases: they could be defined by any known analytic expression in the entire domain of definition, or be represented by different analytic expressions for different values of the argument, they might be continuous or discontinuous, or even assume infinite values, etc., as Fourier more than once points out. But the number of such specific problems is unimaginable, and therefore to limit the character of functional dependence in any way, thereby limiting the generality of the integrals of differential equations, would be to close oneself off from the route to the study of all possible natural phenomena by mathematical analysis, and Fourier was convinced of the comprehensibility of nature; he considered mathematical analysis as broad as nature itself [1, p. xxiii].

---

[36] The reference is to the representation of a function by a Fourier integral.
[37] For another similar statement cf. the book of Rybnikov [1, p. 34].

2.7 The arbitrariness in a functional correspondence

The definitions of the concept of function due to Euler, Condorcet, Lacroix, and Fourier given in the preceding section would seem to leave no doubt that the notion of functional dependence that is now usually called the Dirichlet concept of a function began to form in the middle of the eighteenth century and by 1822 had culminated in the *Théorie analytique de chaleur*. However doubts as to the validity of such an inference do exist, mostly expressed implicitly—the very fact that this definition is attributed to Lobachevskii, Dirichlet or even Bolzano[38] or to the general situation with regard to the scientific activity of these scholars, as in Dieudonné's judgment that Fourier was an eighteenth-century mathematician (in [2, p. 2]). But such doubts are also expressed explicitly. They are perhaps most clearly expressed in the book of Hawkins [1], and on that point we shall go into more detail.

After quoting several places in Fourier's *Théorie analytique de chaleur* where Fourier talks about his understanding of the nature of an arbitrary function, including the part quoted in the preceding section of the present book, Hawkins continues: "These passages, particularly the last sentence above, may suggest that Fourier held an extremely general conception of a function. Actually this is not the case: Fourier's conception of a function was that of his eighteenth-century predecessors." (p. 6). But since Hawkins did not consider the definitions of Euler, Condorcet, and Lacroix given in the preceding section, but gave (pp. 3–4) only Euler's 1748 definition and mentioned that Euler had the concept of a function as a curve drawn by a free movement of the hand (p. 4), he is really asserting that Fourier did not go beyond Euler along these lines. And even after pointing out that Fourier considered functions having a finite number of discontinuities or points of nondifferentiability in the modern sense, Hawkins nevertheless says, "It would seem, therefore, that Fourier had in mind [in the definitions of his that Hawkins cited—F. M.] functions whose graphs are smooth curves except for possibly a finite number of exceptional points" (p. 6).

In favor of such an interpretation of Fourier's understanding of the concept of functional dependence Hawkins gives the following arguments. First, Fourier used the term *discontinuous functions* in the eighteenth-century sense, i.e., to denote functions represented by different analytic expressions in different regions. Second, Fourier is said to have had in mind "a function not given by a single equation" when speaking of an "arbitrary" function. In contrast to his usual practice in this book, Hawkins gives no example of this. Third,

---
[38] Cf. the following section.

Fourier believed that arbitrary functions behave "very well," as confirmed by Fourier's assertion that an arbitrary function is representable by a Fourier integral. The final argument is summed up in Fourier's words on the transform that bears his name: having applied his transformation to the function $f(x)$, Fourier wrote that $f(x)$ "acquires in a manner by this transformation all the properties of trigonometric quantities; differentiations, integrations, and summations of series thus apply to functions in general in the same manner as to exponential trigonometric functions."[39]

Hawkins' arguments as to the concept of a function in the eighteenth and early nineteenth centuries, especially regarding Fourier, cannot be considered convincing. Before considering them in detail we make a general remark.

A distinction should be made between the general meaning a mathematician gives to a general concept he formulates and the concept with which he actually works in his investigations. In order to reason about a concept it does not suffice to understand it in all the generality conceivable in the corresponding period. It must also be subject to the rules of reasoning and analytic calculations and expressed in generally accepted mathematical language of the time. It is natural therefore that some mathematicians, for example in the eighteenth century, were able to have a very general concept of a function yet actually work with a smaller class of functions, since the available mathematical tools were inapplicable to functions of a more general type. It would be incorrect to deny that they had a more general notion of function, especially when they explicitly state such a notion, simply because in their works they constantly talk about functions of a comparatively small class. It would be the same as denying that modern mathematics has a general concept of function. Indeed, if we consider the totality of mathematical literature involved not with the definition of the concept of a function but with the actual study and applications of their properties, the most general concept of function in modern mathematical literature is by and large the measurable function. The literature in which nonmeasurable functions are studied, though it exists, is so diluted by the volume of books and articles devoted to measurable functions, and indeed even functions of smaller classes, that it remains unnoticeable against such a background. And this applies not only to the concept of a function and not only in relation to measurability.

We now pass to Hawkins' arguments with regard to Fourier's concept of a function.

The term *discontinuous function* was indeed sometimes used by Fourier in the sense indicated by Hawkins. But what else could he have done? Such

---

[39] We have quoted Fourier from the book of Hawkins [1, p. 6], who took these words from the English translation of Fourier's work.

was the language of the mathematical literature of the time, and the other meaning of this term did not become established until later. This does not mean that Fourier lacked functions that are discontinuous in the sense that the function changes abruptly when the argument changes continuously. Fourier had such functions (Hawkins, as mentioned above, recognizes this), and he specifically constructs analytic representations of them, although they cannot be described by "a free movement of the hand." To be sure, the functions considered by Fourier belong to the comparatively small class of functions that are discontinuous or nondifferentiable at a small number of points, but it would be unreasonable to expect any more from Fourier.

Fourier was far from being the first to introduce functions discontinuous or nondifferentiable at a finite number of points. On the contrary, the whole thrust of his efforts in considering such functions lay in the fact that they could be given a unified mathematical expression—by a trigonometric series, as a result of which the old distinction between discontinuous and continuous functions no longer made sense and if one desired to continue to use the old terms it would be necessary to take the trouble to give them a new meaning. And when Hawkins asserts that in speaking of arbitrary functions Fourier meant "functions not defined by a single equation", he attributes too general a concept of function to Fourier—a function not representable by a trigonometric series.

The third argument also misses its mark. It is possible, using the Fourier integral, to represent very large classes of functions, especially if representability is understood in some broader sense, so that if we considered functions very well behaved when they are representable as Fourier integrals, we would have to include in this class much more "arbitrary" functions than Hawkins believes. The fact that Fourier did not prove his assertion with the rigor considered desirable today—which, incidentally, he could not have done—does not detract from the correctness of his assertions within very wide limits and *a fortiori* does not mean that he had a narrow concept of a function.

Fourier is also correct in his words cited by Hawkins as a fourth argument, since they can be interpreted in the sense that it was not "arbitrary" functions that were meant, but rather their images under the Fourier transform. Such images have simpler properties than their preimages, and the usual operations of analysis can indeed be carried out on them. And if we take account of modern notions, for example the ideas of L. Schwartz on the Fourier transform, the concluding words of Fourier can also be taken to refer to the preimages.

In saying this we do not wish to give the impression that Fourier anticipated much of the twentieth-century theory of functions and even functional analysis. It is only now possible to read into his writings certain hints of modern views, such as the representability of an arbitrary measurable function by a Fourier series, or the idea of a distribution. Moreover these hints could not have been developed in his time due to the undeveloped mathematical apparatus of the early nineteenth century. But in regard to the concept of a function of a real variable the matter is somewhat more complicated.

It is difficult to say what degree of "arbitrariness" in a functional correspondence Fourier actually had in mind in his writings. If we compare these writings with the "Dirichlet definition," we cannot help noticing not only their external similarity, but also that Fourier's definition, taken literally, is significantly more general than that of Dirichlet (cf. Sec. 8 below). However, the fact that for Fourier the values of a function succeed one another in a completely arbitrary manner and each of them is defined as if it were the only one, that these values are not subject to any single law (incidentally these last words of Fourier were omitted by Hawkins), all this sounds too modern to be understood literally and from the present-day point of view. One has only to ask the following questions—Is it conceivable that Fourier, in speaking of the arbitrariness of a correspondence, could have thought of this correspondence as discontinuous at every point or imagined a correspondence in which the function assumed infinite values on a set of positive measure?—to give a negative answer. Even the question whether Fourier would have admitted the presence of a simple infinity of discontinuities or points of nondifferentiability in his "arbitrary" correspondence clearly must be answered in the negative.

In his way it might seem that we have arrived at Hawkins' conclusion that for Fourier functions were piecewise smooth curves except for a finite number of points of discontinuity, the more so as these are the functions he works with, and he never imagined more general functional correspondences.

It is nevertheless difficult to agree with this assessment, since a similar inference would characterize only half or less of the actual state of affairs. The reason is that the general concepts introduced by mathematicians are almost always rather vague and elusive, and only gradually acquire more or less sharp boundary lines, and the more general the concept the greater its vagueness and the longer that vagueness persists. This is not a defect, rather a virtue of general concepts. It is the vagueness and flexibility of new concepts that opens a wide field of applications to them. Such was precisely the case with Fourier's definition of a functional correspondence. If one may assuredly give negative answers to the questions posed above, one must evidently answer positively such hypothetical questions as, "Would Fourier have considered the Dirichlet function as falling under his definition of a function if it had been shown to him? Would Fourier have recognized a continuous function having a derivative nowhere if such a function had been constructed in 1822?" This is confirmed by the practice of the first half of the nineteenth century, when such objects were introduced into mathematics.

Hence we can answer the question of the arbitrariness in Fourier's functional correspondence by adding the following to what was said above: It was not rigidly defined, with some fixed boundaries, and when necessary it could be broadened and deepened. This applies not only to Fourier but also to Euler.

Indeed it again seems clear that in 1748 Euler adheres mostly to the view of a function as an analytic expression or a single formula. In 1755 he gives the definition of the general concept of a function. But what does he mean by the "extremely broad character" of a functional correspondence, or "all the ways in which one quantity can be determined using others"? It is probably more difficult to answer this than the analogous question regarding Fourier's formulation. One has only to ask whether Euler admitted functions having finite discontinuities to obtain arguments in favor of the response "Yes and no." His curves "drawn by a free movement of the hand" are continuous in our sense, and one can say rather definitely that in the course of his investigations more general functions do not generally occur. But these functions can have derivatives with a finite number of discontinuities at any finite number of points, and Euler talked about such derivatives. Did he not include these derivatives in the class of objects characterized by the term *function*? After all, they cannot be drawn by a free movement of the hand if this motion is to be continuous.

The competition announced in 1787 by the Petersburg Academy of Sciences is also typical. The famous argument over the vibrating string posed in sharp relief the problem of interpreting the "arbitrary function" occurring in an integral of a partial differential equation. Up to 1787 it had been far from a solution, despite the efforts of prominent analysts. The choice of topic and one of the proposed solutions are of sufficient interest to be worth a pause. The topic was posed as follows:[40]

> Determine whether the arbitrary functions introduced in integrating differential equations containing more than two variables belonging to arbitrary curves or surfaces are algebraic, transcendental, or mechanical, whether they are discontinuous or can be obtained by a free movement of the hand; or whether only continuous curves expressible by algebraic or transcendental equations can legitimately be applied.

Here the arbitrariness in a functional relation is thought of as no more general than a curve drawn by a free movement of the hand. But in Arbogast's *Mémoire sur la nature des fonctions quelconques, qui entrent dans*

---
[40] Quoted by Timchenko [1, p. 479].

les intégrales des équations à dérivées partielles, which won a prize from the Academy in 1790, this arbitrariness is significantly enlarged.[41] He shows first that if a function is formed by parts of different analytic curves, then discontinuities of its derivative (in the modern sense) do not prevent the function from satisfying the suggested partial differential equation. In this Arbogast had not yet parted company with his predecessors, since functions with derivatives discontinuous in our sense were admitted by d'Alembert,[42] Laplace, and Condorcet.[43] What is curious in this—and this applies especially to d'Alembert—is that the derivatives of such functions were seemingly not themselves functions.

But Arbogast went further, asserting that even functions that are discontinuous in our sense could occur in the integrals of partial differential equations, and even proposed a special term for them.[44]

A great deal could be written on this topic and the question would still not be settled. As Timchenko correctly remarked, "The mathematicians of the last century [i.e., the eighteenth—F. M.] did not distinguish all the possible peculiarities in the nature of arbitrary functions that are studied with such attention and care in our day. That is why their reasoning is always somewhat vague, and cannot easily be expressed in terms acceptable to modern mathematicians ([1, p. 491, footnote 2]). To this we may add that what Timchenko said applies not only to the eighteenth-century mathematicians.

2.8 The Lobachevskii-Dirichlet definition

As Yushkevich has correctly noted [4, p. 149], the definition of a function proposed by Euler in 1755 was in some respects in advance of the real needs of the time. Moreover both then and now there were no mathematical methods for studying so general a concept. Nevertheless it was not without its uses even in the eighteenth century. One might even say that it was necessary. In any case it was useful.

The fundamental physico-mathematical regularities that were studied in the eighteenth century were described by various differential equations, frequently involving partial derivatives. This is particularly true of the various vibration problems. The integrals of these equations contained an arbitrary function; and though the character of the arbitrariness was not yet quite clear,

---

[41] Here and below we use the book of Timchenko [1, pp. 479–493].
[42] Ibid., p. 486, Footnote 1.
[43] Ibid., p. 487.
[44] Ibid., p. 493.

as became especially clear in the famous polemic on the vibrating string, still the need for sufficient generality in this arbitrariness appeared quite clearly. Therefore a mathematical description of the physico-mechanical notions, as they took shape in the eighteenth century and especially in the nineteenth century, required a concept of function whose generality, and to some degree indefiniteness, would allow the concept of functional dependence to reflect the concept of a dynamical law of nature, which, though it seems not to have been stated in a general form as yet, was nevertheless present in the wide-ranging mind of the scientist.

But independently of this requirement of natural science the Euler definition was needed in mathematics itself. Even in the eighteenth century the theory of differential equations, for example, had reached a very high level of development, and in this theory, considered by itself independently of its connections with the real world, an extremely general concept of function was needed.

Beginning with Hankel [2, p. 49], mathematicians have frequently referred to the Euler definition as nominal definition,[45] giving this predicate a negative connotation and even considering it a fundamental flaw in a definition (Yushkevich [4, p. 149]). One cannot agree with this judgment.

There are indeed, as we shall see below, great difficulties hidden in this definition. But "despite all this, the definition of Euler played [and, we add, continues to play—F. M.] an enormous positive role, having served, if we may use words spoken in another connection, as a 'medium of spontaneous generation' for all the complex and daring constructions of the theory of functions in the second half of the nineteenth century and the first half of the twentieth" (Yushkevich [4, p. 149]). In particular, in he first half of the nineteenth century it served as a "medium of spontaneous generation" for the concept of a continuous function.

It can be assumed that if it had not been for the Euler definition, given widespread publicity by Lacroix, we would very likely not have had the concept of a continuous function due to Bolzano and Cauchy. Indeed, reading the works of Bolzano [1] or Cauchy [1], one naturally notices that they give the definition of a continuous function, but the former does not give the definition of a function in general[46] and the latter introduces it in such a cursory manner [1, p. 18] that it seems no one noticed the general definition of it until Lebesgue [20, p. 4]. But then to what general definition did they refer their special case?

---

[45] Hankel called it the Dirichlet definition, about which we shall have more to say below.

[46] Kol'man [1, p. 53] asserts that Bolzano did state the definition in [1]; we have not been able to find it.

2.8 The Lobachevskii-Dirichlet definition

It is difficult to believe that Bolzano or Cauchy could have been unaware of such fundamental works as Euler's *Differential Calculus*, Lacroix' *Traité*, or the abridged courses of the latter. In fact in 1817 one of the latter (in its second edition of 1806) was translated into German and Bolzano gives a direct reference to it later (Kol'man [1, p. 63]). And in speaking of functions in general, distinguishing the class of continuous functions among them, it was the Euler definition that they consciously or unconsciously used, locating the more tangible class of functions[47] in this "medium of spontaneous generation."

The demarcation of the class of continuous functions was an important epoch in the development of analysis. The stock of Lagrangian "analytic" functions of a real variable had become inadequate even in the eighteenth century. Moreover their properties had by and large all been studied. Continuous functions, in contrast, turned out to be so inexhaustible in their properties that the study of them continues even today. In relation to them it became possible to approach afresh the concepts of integral and derivative, series expansion, etc., in short, to come to grips with the theory of functions of a real variable.

In the Bolzano-Cauchy definition of a continuous function nothing is said about the way in which the functional dependence is defined. But the tradition of expressing the latter as an analytic dependence defined by a comparatively small class of formulas continued to thrive.[48] In addition it remained the most important method of studying functions, although the mere definition of continuity opened new perspectives. But to take better advantage of this opportunity needed to free the concept of a continuous function of its obligatory connection with the old analytic expression. It seems to us that it was this property that was the main peculiarity of the Lobachevskii-Dirichlet definition.

It is nearly universally accepted that Lobachevskii and Dirichlet pro-

---

[47] We leave aside the question of the mutual relationship of Bolzano and Cauchy, in particular in connection with the introduction of the concept of a continuous function. On this question we refer to the articles of Grattan-Guinness [1] and Freudenthal [1].

[48] It suffices to recall that the fourth edition of Lagrange's *Théorie des fonctions analytiques* appeared in 1813. But the stock of functions continuous in the new sense yet not continuous in the old sense was so small that it enabled Grattan-Guinness to conjecture that when Cauchy introduced the definition of a continuous function in 1821 he still meant continuity in the sense of Euler [2, p. 50]. Just how vigorous the old notion of continuity was can be shown, for example, by the following facts: Bunyakovskii, giving the Cauchy definition of continuity in 1839, "supplements it with the hypothesis that the function be representable by a single analytic expression" (Yushkevich [5, p. 298]); in 1847 Stokes describes the two meanings of the term *continuity* as equally valid [1, p. 240], etc.

posed the general definition of a function of one variable as an arbitrary single-valued correspondence of two sets of numbers. Actually the situation with regard to their definitions is more complicated than that, and we shall give some details.

In 1834, in his paper "On the vanishing of trigonometric series" Lobachevskii wrote:

> The general concept of a function requires that a function of x be defined as a number given for each x and varying gradually with x. The value of the function can be given either by an analytic expression, or by a condition that provides a means of examining all numbers and choosing one of them; or finally, the dependence may exist and remain unknown. For example x^3 is a function of x that is expressed analytically; but the root of an equation of degree five is a function of the last term for which no analytic expression has been found and which is determined by the equation as a condition... Finally the conditions to which a function is subject may be as yet unknown, while the fact of the dependence of the numbers undoubtedly exists already. In such a case, the assumption that the function can be expressed analytically must be called arbitrary. It is true that no examples have yet been encountered in which the dependence of the numbers cannot be represented directly or hypothetically by an analytic expression; however, we should not be positive that another assumption would not lead to a new solution... It seems impossible to doubt either that everything in the universe can be represented by numbers or that every change and relation in it can be represented by an analytic expression. *At the same time, a broad view of the theory admits the existence of a dependence in the mere sense of regarding numbers that have some connection with each other as jointly given.* Thus in his *Calcul des fonctions*, with which he tried to replace the differential calculus, Lagrange damaged the scope of the concept by the same amount that he thought to gain in rigor of argument.
>
> Thus, under the name of function we must understand in general a number whose gradual variations are known and depend on the variations of another, even though it be in a completely unknown manner [1, pp. 43–44].

This long quotation nearly coincides with that given by Lunts [2, p. 15] in analyzing the Lobachevskii concept of a function; we have added only the two sentences giving the example of x^3 and the equation of degree five and the last sentence, which were needed for the argument we are about to give.

2.8 The Lobachevskii-Dirichlet definition

The emphasis is missing from the original text; we follow Lunts in adding it.

Lunts deduces from this excerpt that Lobachevskii here gives a clear definition of the general concept of a function. We shall allow ourselves to give Lunts' argument in his own words.

> A comparison of the excerpt we have given with the whole corpus of Lobachevskii's work in mathematical analysis leaves no doubt that the sentence we have emphasized plays the role of a definition of the concept of a function. One can sometimes hear the opinion that Lobachevskii connected the concept of a function with continuity. The argument given in support of this assertion is the first sentence of the excerpt just quoted. The argument is that Lobachevskii's term *gradually* is equivalent to the modern term *continuously*. It should be noted, however, that a strict definition of this term is introduced only in the paper "A method of verifying...."[49] although in introducing that term Lobachevskii himself erroneously claims that the corresponding definition was formulated by him in the paper "On the vanishing of trigonometric series." To be sure, in [1] Lobachevskii occasionally uses the word "gradually" without defining it. However, when Lobachevskii says that a function is a number that "gradually changes with x," he means only that the function varies *to the extent* that x varies [the emphasis is Lunts'— F. M.]. Lobachevskii also uses the word *gradually* in approximately the same sense at the end of p. 44 and on p. 45. The fact that Lobachevskii did not connect the concept of a function with continuity is indisputable, if only because in many places he specifies whether a function is "gradual" or not, studies the convergence of a Fourier series at points where "gradualness" is violated, etc. We must therefore consider it indisputably settled that Lobachevskii has priority for the modern definition of a function (the similar definition of Dirichlet dates to 1837) [2, p. 15–16].

We nevertheless adhere to the opinion that the definition given by Lobachevskii was the definition of a continuous function. In the first place the definition Lunts is seeking in Lobachevskii's writing actually existed long before 1834, as we have seen, and it could not have been unknown to Lobachevskii. Though one may only conjecture that he knew of Lacroix' definition, based on the example of the fifth-degree equation, which appears in both works, there is no doubt in relation to Fourier's formulation—Lobachevskii had definitely read the latter.

---

[49] I.e., in Lobachevskii's paper [2, pp. 130–131].

Second, it follows immediately from the Lobachevskii quotation that it is not the emphasized words that play the role of a definition, as Lunts asserts, but the first sentence. Otherwise Lobachevskii would have to be accused of not observing the rules of grammar. This is confirmed in the last sentence, which Lunts omits.

Third, in the following year he defines the word *gradual* in exactly the sense of the modern word *continuous* [2, pp. 130—131] and at the same time mentions that this is the sense in which he used it in [1], so that there can be no grounds for interpreting it in the sense of Lunts. Moreover in some other places in [1] Lobachevskii absolutely did use the term *gradual* in the sense of *continuous* [1, p. 49], as Lunts himself admits in the last of his remarks on that page.

The fact that Lobachevskii knew about the existence of discontinuous functions is beyond doubt, since he refers directly [1, p. 31] to the paper of Dirichlet [1], which contains not only a proof that a quite extensive class of discontinuous functions can be expanded in Fourier series but also Dirichlet's famous example of a function discontinuous at every point. It is for that reason that he specifies that a function is "gradual" at the point in question and investigates the convergence of the Fourier series at points where gradualness is violated.[50] Thus it is true "that Lobachevskii did not connect the concept of a function with continuity." It is not true, however, that Lobachevskii defined an arbitrary function, since such a definition already existed; he defined an arbitrary continuous function, and his insistent emphasis on the inessentiality of an analytic expression referred to a continuous dependence.

Fourth and finally, the vehement insistence on Lobachevskii's priority over Dirichlet in relation to the definition of an arbitrary (generally discontinuous) function is superfluous, if only because the latter gave no such definition. We now turn to a discussion of this.

Those who speak of Dirichlet's definition refer to his 1837 paper [2]. But here is what we read in that paper in connection with the definition of a function:

> We shall take a and b to be two fixed values and x a variable quantity assuming all values between a and b. If now a unique finite y corresponds to each x, and moreover in such a way that when x ranges continuously over the interval from a to b, $y = f(x)$ also varies continuously (sich ebenfalls allmählich verändert), then y is called a *continuous* (stetige oder continuirliche)[51] function of

---

[50] Having continuity in mind, incidentally, as Lunts also takes for granted in speaking of it.

[51] Since we shall be discussing only continuous functions in what follows, this adjective can be safely omitted [Dirichlet's remark—F. M.].

2.8 The Lobachevskii-Dirichlet definition 61

x for this interval. It is not at all necessary here that y be given in terms of x by one and the same law throughout the entire interval, and it is also not necessary that it be regarded as a dependence expressed using mathematical operations. Geometrically depicted, a continuous function (i.e., one in which x and y are thought of as abscissa and ordinate) is a connected curve on which only one point corresponds to each abscissa between a and b. This definition does not ascribe any law to different parts of the curve; it can be composed of different parts or thought of as not having any law at all. It follows from this that such a function can be regarded as completely defined for an interval if it is either given graphically for the whole interval or subject to different mathematical laws in different parts. If a function is defined for only one part of an interval, the method of continuing it to the rest of the interval is completely arbitrary [2, pp. 133–134].

There is no more general concept of a function in Dirichlet's paper [2]. Moreover, this entire paper involves only continuous functions, as the author warns in the footnote. Thus the entire thrust of Dirichlet's definition is directed, as with Lobachevskii, against the obligatory analytic definition of continuous functions (the possibility of a geometric definition is not excluded, but on the contrary, is assumed) and toward the emphasis on the single-valuedness of the correspondence, which is less marked with Lobachevskii.

Did Dirichlet know about discontinuous functions? Absolutely! As we mentioned, in his paper [1], published eight years earlier, he had not only proved that a large class of discontinuous functions could be expanded in Fourier series (those having a finite number of discontinuities) but had also proposed an example of a function discontinuous everywhere. In addition, in the same paper he had attempted to extend the definition of the integral to certain functions with an everywhere dense set of discontinuities.[52] (In the 1837 paper he limited himself to integrating continuous functions.) The footnote in the quotation from Dirichlet shows explicitly that he deliberately limits himself to continuous functions.

It has already been noted that it was the definition of a continuous function that Dirichlet gave in [2]. This was pointed out, for example, by Church [1, p. 22], and later by Steiner [1, pp. 18–19]. Others, no doubt, have also known about this. But the position of those who have spoken about Dirichlet's definition has generally been expressed as follows:

> It is not important that Dirichlet restricts his statement at this particular place to continuous functions, since it is clear from other

---

[52] On this see Hawkins [1, p. 13].

passages in his writings that the same generality is allowed to discontinuous functions (Church, [1, p. 22]).

This restriction seems important to us, however. As we have already said, a general definition of a function had long since been in existence. Dirichlet, like Lobachevskii, could not have failed to know of it. Even allowing that Fourier had given his definition (given above in Sec. 7) a less general sense than Dirichlet would have given it (and this is quite likely), it is nevertheless true that if we ascribe to the latter the plan of a general concept of function of one real variable, we must explain why, demonstrably knowing about the repeated statements by Fourier regarding such functions, he never once mentioned them. It must also be explained why no one seems to have claimed that this definition was due to Dirichlet until Hankel did so in 1870. If indeed such a general concept dated only to 1837, Stokes in 1847 [1, pp. 239–240] and Riemann in 1851 [1, p. 1] would hardly have spoken of it without mentioning Dirichlet. They undoubtedly knew not only of Fourier's definition, but also of the earlier definitions, if only from the course of Lacroix, and therefore quite properly did not connect them with anyone's name—definitions that have appeared in standard textbooks for a long time are quite frequently anonymous.

Consequently, like the definition of Lobachevskii, the definition of Dirichlet referred only to continuous functions. What was the reason for introducing it, when the definitions of a continuous function by Bolzano and Cauchy already existed? We have given the reason at the beginning of the present section, and we shall summarize it again as follows: the size of the class of continuous functions was not clear from the definitions of Bolzano and Cauchy; the old meaning of the term "continuity," defining a function by a single formula, remained in existence alongside its new meaning almost until the middle of the nineteenth century, and it was necessary to become free of such an interpretation of continuous dependence and give the newly established term the most general content, since it was not possible at the time to exclude *a priori* the possibility that a function continuous in the Bolzano-Cauchy sense is also analytic in the sense of Lagrange.

The definitions of Lobachevskii and Dirichlet were also intended to avoid the possibility of this mistake, to display the full breadth of precisely the concept of a continuous function. And if we take Luzin's point of view that classical analysis and the theory of functions, though they have the same subject matter—the concept of a function—differ in that the former studies functions defined by an analytic expression while the latter studies functions starting from a general definition of them [4, p. 48], we can say that the theory of functions began with the introduction of the concept of a continuous function, since its properties could be studied on the basis of a general definition, and not an analytical dependence. The Lobachevskii-Dirichlet definition opened the way to the function-theoretic approach to the study of the properties of continuous functions.

It is evident that Hankel was responsible for the fact that up to now the general concept of a function of one variable is mostly connected with the name of Dirichlet. In 1870 he brought into general use the phrase "definition of a function according to Dirichlet," meaning just such arbitrary functions [2, p. 49], and this seems rather surprising, if we consider Hankel's constant interest in historico-mathematical studies. The explanation of his mistake (and that of many mathematicians after him) can be found in the fact that the earlier definition of Euler (even as expounded by Lacroix and amended by Fourier) was by then far off in time, and mainly in the fact that since the work of Riemann [2] really arbitrary functions, not subject to any analytic laws, had become an object of active investigation. And Hankel, thinking of an arbitrary correspondence and knowing of the existence of discontinuous functions in the investigations of Dirichlet, was able to read the general definition of a function into Dirichlet's definition.

2.9 The extension and enrichment of the concept of a function in the nineteenth century

The pretentious title of the present section would require the writing of a whole series of books to reveal its full import. Some of these books have already been written, for example by Markushevich [1], Paplauskas [1], Rosenthal [1], Grattan-Guinness [2], and Hawkins [1]. Without making any claim of completeness we shall try only to summarize some particular results connected with the extension and enrichment of the concept of a function.

Clearly the main achievement of the nineteenth century in this respect was the creation of the theory of functions of a complex variable. Referring the reader to the book of Markushevich [1] in this connection, we note merely that the earlier definitions were significantly extended in that correspondences between sets of complex and real numbers (real-valued functions of a complex variable) and complex number-to-complex number correspondences (complex-valued functions of a complex variable) began to be considered instead of single-valued correspondences between sets of real numbers. When this was done—and this is also a step away from the definitions under consideration— such correspondences turned out in general not to be single-valued. In the theory of functions of a complex variable both particular functions and diverse classes of functions were profoundly studied; new tools were discovered and developed for studying the properties of functions, and the tools that already existed were sharpened. The theory itself originated to a large degree under the influence of the demands of mathematical science, and its results found diverse applications there.

Furthermore, in the definitions of the concept of a function that were given, the topic was functions of one variable. But functions of several variables had been considered long before these definitions.[53] Euler had extended Bernoulli's definition to functions of several variables in 1748 [1, p. 60]; in this same place (Ch. V) he studies their elementary properties. The deeper properties of functions of several variables were studied by many mathematicians of the eighteenth and nineteenth centuries. Unfortunately the history of the analysis of functions of several variables is almost completely undeveloped, and therefore we limit ourselves here to just two questions. First of all, the number-to-number correspondence studied in the theory of functions of one variable (real or complex), is complicated into pair-to-number, triple-to-number, and in general n-tuple-to-number correspondences. In this situation an n-tuple of numbers does not necessarily denote a point in an n-dimensional region in its traditional interpretation as a plane or a rectangular part of it, three-dimensional space or a parallelepipedal part of it, etc., but rather as an n-tuple of numbers belonging to a more complicated manifold. Functions are considered that are defined on curves in two-dimensional space, on surfaces in n-dimensional space, and in general on m-dimensional manifolds in n-dimensional space ($m < n$). To study such functions the corresponding methods of investigation are developed—multiple series, multiple and surface integrals, partial and normal derivatives, and the like.

We note in passing that although the theory of functions of several variables was quite widely developed in the nineteenth century, one can evidently assert nevertheless that one of the distinguishing characteristics of the theory of functions of the last century, in contrast to the present century, is the predominance of the study of functions of one variable.

It must still be mentioned that the extension of the concept of a function in the nineteenth century occurred also in vector analysis in its extension to number-to-vector correspondences (vector-valued functions of a real argument) and vector-to-vector correspondences (vector-valued functions of a vector argument). By the end of the century the ideas of quaternion-valued functions and functions of more general number systems were appearing.

In the theory of functions of one variable itself there occurred a notable extension connected with the creation of set theory. Up to this time functions were regarded as defined either on the entire real line or on an interval of it.

---

[53] As an example we note Newton's investigations of gravitational theory in his *Principia*, where Newton not only applies in essence functions of several variables, but actually develops the elements of integral calculus for them. For this cf. the article of Antropova [1].

2.9 Extension and enrichment in the 19th century

The introduction of point sets made it possible to consider functions defined on sets of a very general type. This idea was promulgated by Dini [4] and later by Pasch [1, 2] with full consciousness of what they were doing and was fruitfully carried out, moreover for functions of several variables, by Jordan [2, 3].

Along with such extensions of the domains of definition of functions there occurred an enrichment of the definitions formulated earlier.

First of all various classes of functions were distinguished and studied. Besides continuous functions, which were discussed above, the classes of functions satisfying a Lipschitz condition (1864), functions integrable in the sense of Riemann (1867), and functions of bounded variation (Jordan, 1881) were introduced, to name only the most important.

These extensions were followed by continuous progress in sharpening a large number of definitions connected with the concept of function, and new definitions were introduced. At the very beginning of the nineteenth century in the writings of Lagrange or Lacroix the continuity and differentiability of a function are inseparably linked, and many mathematicians "prove" that every continuous function is differentiable except possibly at a finite set of points.[54] But as early as 1830 Bolzano, in a manuscript not published until a century later, not only recognizes that these are different concepts having a comparatively weak connection with each other, but even constructs an example of a continuous function that is nondifferentiable on an everywhere dense set of points.[55] In the years 1834–1835 Lobachevskii makes a distinction between the concepts of continuity and differentiability, though he assumes that a continuous function can have only isolated points of nondifferentiability.[56] However, after the publication of of the works of Riemann [2] and Hankel [2] more and more particular continuous functions begin to appear, having no derivative on the most diverse kinds of sets, even the points of a whole interval (Weierstrass [5], first published in 1875 by Du Bois-Reymond [3]). Such investigations were later continued by many people. Besides the numerous examples of individual continuous functions without derivatives infinite classes of such functions were constructed (Dini [1–4], Darboux [2]); moreover in 1931 S. Mazurkiewicz [2] and Banach [1] proved that the set of continuous functions having a derivative at even one point is in a sense negligibly small compared to the set of all continuous functions.[57]

---

[54] Attempts to prove this erroneous proposition continued almost to the end of the 1870's.

[55] For details cf. Bržečka [1], Kol'man [1, pp. 63–70].

[56] Cf. Lunts [2, pp. 16–17].

[57] Of the papers in which the history of the problem of continuous functions without derivatives is traced to a greater or lesser degree, we note the books of Pascal [1, pp. 91–128], Brun-

Ideas about integration were also sharpened. After the Cauchy integral (1823) there appeared the Riemann integral (1867) and it was really with the latter that the set-theoretic method began to be used more and more frequently, first in the theory of integration and then in many other questions of mathematics (Hankel, Smith, Dini, Peano, Jordan, and others). Space does not permit a detailed discussion and therefore we limit ourselves to a single observation.

Throughout the nineteeenth century existence questions played a large role in mathematics. They were solved for the Cauchy and Riemann integrals, regarded as limits of sums. However in reading the works of the mathematicians of the last century one gains the impression that the overwhelming majority of them continued to use the definition of a definite integral due to Newton and Leibniz, in which the definite integral is regarded as the difference of two values of a primitive function. In this connection the question of the existence of a primitive was either not considered at all or was answered by starting from the notion of an integral as a limit of sums, which was logically circular. It was only in 1905 that Lebesgue [11] succeeded in proving the existence of a primitive for every continuous function without resorting to the definition of an integral as a limit of sums; and only then was it justified to regard the integral of a continuous function as the difference of two values of a primitive.

On the other hand, we may note that a similar logical circle arose in certain other questions. Thus, for example, in proving the additivity of the integral, mathematicians usually relied on a theorem about the limit of sums, although for a limiting passage of the type used in the definition of an integral no such theorem had been proved (in the traditional textbooks this theorem was proved only for the limit of a sequence). It became possible to escape from this logical circle only when the concept of a limit over a filter was created and its theory developed. The nineteenth century mathematicians strangely did not notice these kinds of logical insufficiency.

Besides the development of the theory of differentiation and integration, the apparatus of various kinds of series, products, and continued fractions was perfected, various forms of convergence of sequences of functions were introduced, methods of reasoning were sharpened, etc.—in short, the theory of functions was built on a broad base.

---

schvicg [1, pp. 337–340], Pasch [3, pp. 122–129], and Hawkins [1, pp. 42–54], and also the articles of Hardy [4] and Knopp [1].

2.10 The definition of a function according to Dedekind

In the first section of this chapter we gave the general definition of a function that until recently was the most widely accepted, an arbitrary single-valued correspondence between elements of two abstract sets and in the preceding section we attempted to point out the milestones on the road to such an extension of the original definition as an arbitrary correspondence between two sets of numbers or points. It must be admitted, however, that the decisive factor in this transition to a completely abstract argument in the domain and range of a function nevertheless consisted not of the extensions mentioned in the preceding section, but of considerations that specifically go beyond the boundaries of theory of functions.

For the argument and the value of a function to be regarded as elements of abstract sets it is necessary that such sets become an object of study. They became such in the abstract set theory created by Cantor, Dedekind, and others. Only after the elements of such a theory were constructed, or to be more precise, in the course of the construction, did it become possible to give the concept such a general character. Evidently this definition of a function was first stated by Dedekind [5] in 1887. Indeed, after introducing a completely general concept of a set (a system, in Dedekind's terminology) [5, pp. 1–2], he wrote:

> By a *transformation* φ of a system S we shall mean a law by which to each definite element s of the system there belongs a completely definite thing called the *image* of s and denoted by the symbol $\varphi(s)$; the same situation can be expressed in various ways: $\varphi(s)$ corresponds to the element s, or $\varphi(s)$ is obtained from s by the transformation φ, or s goes into $\varphi(s)$ by the transformation φ.
> If now T is some portion of S, then of course a transformation φ of the system S must contain some transformation of the system T, which for simplicity may be denoted by the same symbol φ, and the meaning of this symbol is that to each element t of the system T there corresponds the same image $\varphi(t)$ that t possesses when regarded as an element of the system S; the combined system consisting of all the images $\varphi(t)$ can be called the image of T and denoted by the symbol $\varphi(T)$, thus explaining the meaning of the symbol $\varphi(S)$ [5, p. 5].

How generally Dedekind conceived of this law or transformation can be seen if only from the fact that he immediately gives as an example of a transformation the mere naming or denotation of the elements of the set under consideration with definite signs. He considered the possibility of establishing a correspondence between the elements of sets to be a special ability of the human mind, without which "all thought is impossible" [5, pp. iii–iv].

The transformation φ of a given set S may lead to elements of a different set Z or may give elements of the same set S, and then Z is a subset of S and φ is called a transformation of S into itself [5, pp. 8–9].

It may be that the most interesting thing from the historical point of view is that in introducing the general concept of a transformation Dedekind did not point out its connections with the traditional concept of a function of mathematical analysis, although these connections can be implicitly traced in his book. Apparently this concept did not form in his mind in the way one might assume on the basis of the preceding section: there was a general concept of a function as a transformation of one set of real numbers into another; it gradually broadened to include correspondences between different, not necessarily numerical sets; after the introduction of the concept of an abstract set, the preceding extensions attained their apogee in the statement of the definition given above.

Judging from the fact that after introducing this definition and studying its simplest properties Dedekind [5, pp. 5–6] proceeds to study the one-to-one correspondence of sets (pp. 7–8), one might conjecture that the latter was one of the sources of the generalization. Indeed, in the Cantorian one-to-one correspondence of two abstract sets, which he had studied as early as 1878 in connection with the general notion of cardinality (G. Cantor [4, p. 119]), we have the special case of the definition that Dedekind calls a *similarity transformation* and Bourbaki calls *bijective* [1, p. 94]). However, the actual picture is somewhat more complicated.

The fact is that Dedekind had used the concept of a transformation in his number-theoretic and algebraic investigations considerably earlier. Thus for example, as early as 1871 he had studied a transformation of one algebraic field into another [1, pp. 223–224], and in a paper of 1877 [4] we can even read between the lines the general definition given above. Indeed, at the beginning of § 16 we read: "By substitution is usually meant the act by which objects of study or elements of investigation are replaced by corresponding objects or elements and the old elements are said to be replaced by the new elements through this substitution" [4, p. 151]. In content this formulation differs from that of 1887 only in using a different terminology and by less clarity in the terms themselves. But that Dedekind thought of the term *substitution* as a correspondence is confirmed by one of his earlier statements: "...if to each number a of the field A there corresponds some number $b = \varphi(a)$, so that $\varphi(a+a') = \varphi(a) + \varphi(a')$ and $\varphi(aa') = \varphi(a)\varphi(a')$, then the numbers b (if they are not all zero) also form a field $B = \varphi(A)$, which we call *conjugate* with A

2.10 The Dedekind definition of a function

and which is obtained from A by the substitution φ" [1, pp. 424–425].[58]

In both [1] and [4] Dedekind applies this last definition to study the special case of fields of algebraic numbers; for them, incidentally, he introduces the concept of a one-to-one correspondence ([4, p. 152] of the first edition), and also the term *correspondence*. He calls fields that are equivalent in this sense *conjugate*. But if we take into account Dedekind's assertion [5, p. iv] that the plan for the 1887 work occurred to him before 1872 and that he wrote a draft of the book [5] between 1872 and 1878, with which several mathematicians became acquainted at that time, it may be that the Cantorian one-to-one correspondence in its general form owes something to Dedekind's ideas, since Cantor was very likely among the mathematicians who read the draft.[59]

Thus one can assert with a certain degree of assurance that the definition of a function or a transformation in the writing of Dedekind has as its basic sources number-theoretic, algebraic, and set-theoretic investigations. This of course does not mean that the concept of a function that formed in analysis and the theory of functions had no influence on the origin of the abstract notion of a function. The latter disciplines were fundamental divisions of the mathematics of the nineteenth century, and the ideas that grew in them, in particular the idea of functional dependence, could not be without effect on these definitions, as is lucidly shown by Dedekind's notation alone.

Several years later Cantor arrived at the same general concept of a function as Dedekind. In Cantor's work [8], published in 1895–1897, we read:

> By a *covering* of a set M on the elements of a set N, or more simply, a covering of M on N, we understand a law by which with each element n of N there is connected a definite element of M, and the same element of M may be repeated. The element of M connected with n is in some sense a single-valued function of n and may be denoted, for example, by $f(n)$; it is called a covering function; the corresponding covering of the set N will be denoted $f(N)$ [8, p. 287].

Having introduced this definition, Cantor did not point out that Dedekind had actually had it earlier, and it was sometimes connected with Cantor's name alone. Without denying Cantor's role in forming this definition, we are nevertheless inclined to connect it with the name of Dedekind.[60]

As noted earlier (Sec. 1 above) two undefined concepts are apparent in this definition—set and transformation (correspondence). Moreover the statement of the definition contains propositions: "a transformation is a law," "a

---

[58] Page numbers refer to the original 1871 edition of Dedekind's book.

[59] At the time there was an intensive exchange of opinions on various questions of set theory.

[60] Cf. also Steiner [1, pp. 25–26].

transformation is a correspondence" (or, in other formulations, "a transformation is a rule"). Stated in this form, these propositions contain a logical circle. This defect of the definition was noted in 1911 by Peano [4], who considered the words *correspondence, relation, operation*, and the like to be synonyms. In order to decrease the number of undefined terms and at the same time avoid an explicit vicious circle he proposed a definition of a function that relied on the single undefined term *set*—the second definition we considered in the first section of this chapter, referring to the books of Kuratowski and Mostowski [1], Mal'cev [1], and Shikhanovich [1]. We did not give Peano's own statement, since his symbolism would have required lengthy explanations.

2.11 Approaches to the concept of a function from mathematical logic

Scholars arrived at an equally general concept of function from a somewhat different direction—through the studies of mathematical logic by Frege, Peirce, Schröder, Russell, and others. Here we find ourselves in a more difficult position than in the preceding case, since the history of mathematical logic has not been our specialty; therefore we shall have to make more frequent reference to the historical studies of other authors rather than to the original sources, mainly to the works of Biryukov [1], Styazhkin [1], and Steiner [1].

Ignoring the concept of a function in the writings of Boole, De Morgan, and other predecessors of Frege,[61] we begin by considering the views of the latter. They are expounded by him mainly in his book "Funktion und Begriff" [1], published in 1891.

The first thing to be noted in connection with Frege's concept of a function is the incorrect emphasis many authors lay on his being the first to give the argument and values of a function the generality that is accepted in modern mathematics.[62] This important step in the development of the concept of a function had been taken earlier, as we have seen, by Dedekind. His concept of a transformation (function) is so general that it accommodates all the enlargements of the meaning of the argument and domain of definition spoken of by Frege, so that the content of the concept of function in this sense is the same for the two authors. The following circumstance is of interest in this connection.

It was pointed out above that the mathematician Dedekind, in broadening the domain of definition for the argument and values of a function, made

---

[61] For some data on this see the book of Styazhkin [1, pp. 323, 249, 431, and elsewhere].

[62] Cf., for example, Church [1, p. 23], Biryukov [1, p. 152], Styazhkin [1, p. 431].

2.11 Approaches from mathematical logic

no direct reference to the enlargement of them mentioned in Sec. 9 above. The logician Frege, in contrast, approaching this notion of a function, refers to several of the extensions pointed out [1, pp. 23–24]. In addition he even digresses into the history of the concept of a function in mathematics.

One also cannot agree with Church that it fell to the lot of Frege also to take the important step consisting of "the elimination of the dubious notion of a variable quantity in favor of the notion of the variable as a kind of symbol" [1, p. 23].[63] Church himself notes (p. 23) that even in the Eulerian definition of a function as an analytic expression Euler's *variable quantity* looks like Frege's concept of a variable, although he specifies here that such an interpretation of Euler's understanding of this term is inconsistent with the context of his other statements. But in the case of Euler there is no need to resort to any indirect considerations, since he himself explained his understanding and we have given the corresponding words above in Sec. 5.

To be sure, just as the character of the arbitrariness in a functional correspondence has been differently interpreted at different stages of mathematical development, so the generality of the allowable values of an argument and functions and the distinctness of the meaning of a variable quantity have changed with time, becoming richer and broader. Of course Euler, in explaining his understanding of the term *variable* did not have in mind the generality attained by Frege—a century and a half of mathematical thought were needed for that. In the first place even Dedekind did not imagine all the extensions of mathematical logic spoken of by Frege, though it is curious that one of Frege's most general examples of a functional correspondence—the capital of a country [1, p. 27]—resonates with Dedekind's explanation that the concept of a transformation includes even the mere naming of objects.[64] In particular it can hardly be said that Dedekind somehow connected his notions of transformations with the logical or propositional functions that occupy such an important place in Frege's system.[65]

In our opinion Frege's principal merit in the development of the concept of a function is the following circumstance. As noted above, Dedekind defined the concept of a function in terms of two undefined concepts: set and correspondence, and Peano reduced the number of undefined terms to one—that of set. But the latter is extraordinarily complicated and has not been clarified even now. Space does not permit us to dwell on this, in particular on the paradoxes of set theory. Nevertheless the desire to avoid the rather unclear

---

[63] Cf. also Biryukov, [1, p. 150].

[64] Evidently Frege was already aware of Dedekind's work [5] by 1891, although there are no direct references to it in his article [1].

[65] For this cf. Biryukov [1, pp. 151–152].

notion of a set in defining so important a mathematical concept as function is legitimate. It was such an attempt that Frege made, and this attempt is the more remarkable in that the paradoxes of set theory themselves had not yet been discovered, so that there was apparently no reason for regarding with suspicion the set-theoretic approach to the definition of a function, which had justified itself in so many questions.

Frege's approach was radically different from all those considered above: he introduced the concept of a function without any definition and explained it only through descriptions, historical digressions, and examples.

He had good reason for doing so. Indeed, in mathematics, even if the concept of a set was considered undefined, it had nevertheless been accepted since the work of Cantor that a set is defined when it was possible to decide whether any given object of thought belongs to the set or not; and Dedekind himself had written frankly that a set S is completely defined only when "it is known with regard to anything whether it is an element of S or not" [5, pp. 1–2].

The sets in a Dedekind transformation are defined, and this means that each element of them is endowed with some characteristic property or criterion by which its membership in the set can be decided. But this "endowment with a criterion" falls under the general notion of a Dedekind transformation, and we again encounter a logical circle of a somewhat different character from that spoken of before the Peano definition; a transformation is introduced by a set that itself is characterized by a transformation.

Evidently following approximately this line of thought, Frege took the concept of a function as a primitive undefined concept and defined others in terms of it, including the concept of a relation.[66]

Other important ideas of Frege related to the concept of a function are, in our view, his emphasis on the indissolubility of the connection between the argument and the set of values of the function, their wholeness [1, pp. 19–20], and his critique of a variety of erroneous notions of mathematicians [1, pp. 17–18]. This critique is developed by him in more detail in the 1904 paper [2].

In contrast to Frege, in 1895 Schröder [1, p. 553] took as a starting concept the concept of a relation, and regarded a transformation (function) as a single-valued binary relation, so that he is probably the one to whom the definition in the first section of this essay with the reference to the work of Tarski [1] is due.

Russell expounded his views on the nature of the concept of a func-

---

[66] For more details cf. Biryukov [1, pp. 145–149], Steiner [1, pp. 34–38]. In 1904 Frege [2, p. 88, footnote] pointed out the vicious circle in Hankel's definition.

tion with the greatest clarity, evidently, in the article "Sur la rélation des mathématiques à la logistique" [2]. He takes as undefined the concept of a statement or a propositional function with one, two, or more variables (p. 907). A relation is defined by him as a propositional function of two variables (p. 908). A function is for him a single-valued relation (loc. cit.).

If we take into account that another undefined term *variable* occurs in this chain of definitions,[67] then after working out some details (replacing *variable* by a set and defining a propositional function of two variables in terms of a propositional function of one variable—which Russell did not do in either [1] or [2], and for which the concept of the Cartesian product of two sets is needed), we arrive at the definition given in the first section of this essay with the reference to the book of Kuratowski [2].[68]

We have thus traced the original sources of the definitions of the general concept of a function that we stated at the beginning of this essay. The circle of such statements could be enlarged with a corresponding continuation of the historical survey. But in the last two sections we have already gone beyond the theory of functions proper, and to continue would require an even greater departure from it. In any case, in the theory of functions mathematicians work with a concept of function whose nature, one might say, does not fit into the framework of the seemingly ultimate generality of the definitions considered above. We now turn to the consideration of this new notion of a function.

2.12 Set functions

Here, as above, we shall begin by stating the definition and explaining its relation to our earlier definitions, then pause briefly to give its history. In the book of Natanson [1, Vol. 2, p. 94], for example, we read, "Let $\mathfrak{A}$ be a family of sets e, $\mathfrak{A} = \{e\}$. If to each set $e \in \mathfrak{A}$ there corresponds some number $\Phi(e)$, we say that a *set function* is defined on the family $\mathfrak{A}$."

In all the definitions considered above for the concept of a function the discussion centered on a single-valued correspondence between elements of two sets or on a single-valued relation of elements. In contrast, when we speak of set functions, we mean a set-to-element correspondence. If we follow the letter of the Dedekind definition, for example, it might seem that the concept

---

[67] In [1, pp. 5–6] Russell writes, "The notion of a variable is one of the most difficult with which Logic has to deal, and in the present work a satisfactory theory as to its nature, in spite of many discussions, will hardly be found."

[68] In [1, p. 263] Russell essentially adopts the point of view of Schröder on the interpretation of a function.

of a set function could easily be reduced to the concept of a transformation: it suffices to regard the sets e of the family α as elements of a set M and speak of a correspondence of numbers to the various elements of this set. When mathematicians speak of set functions however, they never seem to mention the possibility of such a reduction, but rather introduce this concept independently (at least in the cases known to me), and this is evidently not coincidental.

The point is that in the theory of set functions, which began to be widely developed only after 1910, it was necessary almost from the beginning to talk about differentiating one set function with respect to another or integrating a point function with respect to a set function. In both cases point functions appear together with set functions. For that reason any attempt to reduce the concept of a set function to the concept of a point function will lead to regarding both the elements of the set e and the set e itself simultaneously as elements of a set M. This, as is known, is forbidden by the Schröder-Russell theory of types[69] because of the appearance of paradoxes in such situations. By 1910 the paradoxes of set theory were sufficiently well known that the fact of their existence influenced the formation of new concepts. It was evidently in order to avoid them that mathematicians consciously or unconsciously made no attempt to reduce the concept of a set function to the concept of a transformation, but regarded the former together with the latter as independent, though connected. We shall adopt this point of view in what follows, regarding the idea of a set function as a new enlargement of the concept of functional dependence considered earlier and not reducing to it.

As in the case of an elementwise functional dependence, the problem of the origin of the concept of a set function can be regarded as having a long evolution. Having at our disposal such a general concept of a set function, as formulated at the beginning of this section (we shall not discuss more general definitions), we can, for example, interpret an integral as a set function,[70], and then the origin of the latter must be assigned to quite remote times. Furthermore, if we regard the length of an interval as a set function (interval function), we could even argue about which came first: the idea of a function or the idea of a set function. In any case prefigurations of set functions can be found, just as was done for prefigurations of point functions, in very ancient

---

[69] The theory of types is usually attributed to Russell. For information on its presence in the writings of Schröder as early as 1890 cf. Biryukov [2, pp. 40–42]. Russell gave the first draft of this theory in 1903 [1, pp. 523–528] without referring to Schröder, although he referred to Schröder many times in this book.

[70] Cf., for example, Lebesgue [20, pp. 143–144].

mathematical investigations. In addition, evidently, one can assert positively that a reasonably conscious introduction of the concept of a point function took place significantly earlier than was the case for the concept of a set function.

We connect the latter for the time being with the name of Cauchy and refer to 1841, the year when his "Mémoire sur le rapport différentiel de deux grandeurs qui varient simultanement" [3] appeared.[71] Cauchy, of course, did not speak of functions of a set or region—he called them coexisting quantities—but the founders of the theory of set functions, Peano [1, p. 167; 5] and Lebesgue [20, p. 290–296], took his reasoning in that sense. We shall not go into detail about the contents of the paper of Cauchy just named, referring the reader to the places in the writings of Peano and Lebesgue, or to our article [1]; instead we note the following.

The introduction of the concept of a set function (coexisting quantities) by Cauchy is inseparably linked with physical and elementary geometrical considerations, which he was using as a guide in an attempt to construct a calculus of set functions. He was constructing this calculus in analogy with the usual differential and integral calculus for point functions, but—and this should be emphasized—he regarded it as a generalization of classical analysis that included the latter as a special case.

Cauchy's ideas did not receive recognition immediately, despite the propaganda given them by Moigno. But starting in the 1870's and 1880's an ever increasing number of mathematicians began to arrive at similar ideas: this was connected with the development of algebra (Dedekind) and especially with the creation of the theory of point sets (Hankel, Cantor, Peano, Pasch, Jordan, and others).

The highest achievement of the nineteenth-century mathematicians in developing the subject of set functions is the fifth chapter of Peano's book [1]. There an even more general concept of such a function was introduced than the one stated at the beginning of this section, since Peano considered it possible to study vector-valued set functions.[72] He introduced the derivative of one set function with respect to another and the integral of a point function with respect to a set function, distinguished the class of additive set functions,

---

[71] If we take into account the fact that Cauchy had expounded these ideas in lectures in the École Polytechnique in 1829–1830, the date for the introduction of set functions moves back to that time.

[72] Not quite in the modern sense, in which the argument of the function ranges over a family of abstract sets and its values are vectors of some functional space; Peano considered only vectors of a finite-dimensional Euclidean space.

and gave examples of nonadditive set functions.[73] It is now possible to state many criticisms of various methods of reasoning applied by Peano [1], but even so in the generality and profundity of its individual ideas this chapter of Peano's book excels Lebesgue's 1910 memoir [16], which is usually considered the source from which sprang the endless stream of modern investigations in the theory of set functions.

Peano himself did not subsequently pursue his investigations in this area, since mathematical logic became his major interest from 1888 on. His investigations also did not find immediate support among his contemporaries. But the theory of set functions continued to form gradually, if only in the investigations of Jordan, Borel, and Lebesgue in measure theory. Without in any way attempting to write the history of the theory of set functions, we note merely that by 1910 the internal mathematical foundation had been created for constructing a more satisfactory theory of these functions, and such a construction was begun by Lebesgue with his memoir "Sur l'intégration des fonctions discontinues" [16]. His studies were soon joined by those of Radon [1], then Vallée-Poussin, whose book [2] played an important role in spreading the ideas of the new mathematical subject, and others. The theory of set functions is now an extremely ramified mathematical discipline with a boundless field of applications.

In constructing the new theory in [16] Lebesgue started from the internal mathematical need to extend his results on the theory of differentiation and integration, which he had developed earlier mainly for the one-dimensional case, to n-dimensional Euclidean space, and seemingly in 1910 had not yet thought of any physical applications of this theory. It was Radon in 1913 who pointed out that mathematical physics cannot get by without the concept of a set function and actually used it [1, p. 1290, footnote]; he himself gave no direct physical applications in [1] unless one counts the indirect applications via the theory of integral equations and the theory of linear operators. The same can be said about his second paper [2] in 1919, although here the connection of the theory of set functions with mathematical physics is more emphasized. By 1928 Lebesgue had become acquainted with Cauchy's paper [3] and it was evidently through that paper that he arrived at the idea of describing the basic physical phenomena not in terms of point functions, but in the language of the theory of set functions as the one best adapted to the needs of physics [20, pp. 290–296].

Not long before this Gyunter had arrived independently at the same concept, and while Lebesgue stated it only in the most general form, Gyunter did much toward restructuring mathematical physics according to such a plan.

---

[73] For more details cf. Medvedev [3].

Of his numerous works devoted to developing a new approach to the mathematical method of describing physical phenomena we limit ourselves to a reference to his works [1] and [2].

Subsequent reactions to this were mixed. But even if we agree that as a whole this approach did not live up to expectations, it was by no means useless. Set functions broke new ground in mathematical physics through probability theory in the form given to it by Kolmogorov, and there can be no doubt that Lebesgue's ideas, and perhaps also those of Gyunter, played a role in this penetration. The penetration was not entirely one-sided, since it was not only the representatives of the theory of set functions who sought applications of the area of mathematics they had developed; the representatives of mathematical science were also striving to find more adequate mathematical means of describing the real-world phenomena they were studying. Science and mathematics were in a way moving toward each other.

2.13 Some other functional correspondences

In studying the concept of a function in the present essay almost all our attention has been concentrated on functional correspondences of Dedekind transformation type, i.e., on elementwise correspondences between two sets; only the preceding section was devoted to set-to-element correspondences. This basically parallels the historical process, if we restrict ourselves to the theory of functions of a real variable in the interpretation we have adopted. But if we keep in mind other mathematical disciplines, especially the theory of functions of a complex variable, the ideas we have been considering will have to be enlarged.

We begin with a general consideration.

Various relations are logically conceivable between sets. In particular the situation of element-to-element transformations and set-to-element transformations suggests an elementary-logic extension to element-to-set and set-to-set transformations. The present section is devoted to showing that such an extension has content. Here we do not claim to be giving even an incomplete survey such as that of the previous section, since all this goes beyond the limits we have marked out.

Let us begin with element-to-set correspondences. If we translate them into the more usual language, they are nothing but multivalued functions. In the formulations given in the first section of this essay it was strongly emphasized that the concept of a function must necessarily accommodate the notion of a single-valued correspondence or relation, i.e., that to a given element of the domain of definition of the function there should correspond only one element of the set of values.

However such a restriction is not justified in logic or history. Indeed, there are no logical grounds for excluding multivalued functions from consideration, and historically they arose in analysis from the time a function was distinguished as an object of analysis. And while mathematicians tried strenuously to avoid them, not always successfully, nevertheless the very fact of their persistence in mathematical investigations forces us to be on guard and to regard this restriction on the concept of a function with some suspicion. Such multivalued functions arose long ago, and, for example, the problem of inverting an ordinary single-valued function leads as a rule to a multivalued function; only in the special case when the functional correspondence is one-to-one will the inverse of a function be single-valued. There are neither logical nor general mathematical grounds for refusing to consider the operation of inversion in its general form, just as there are no grounds for refusing to consider the operation of subtraction when the subtrahend is larger than the minuend. Of course specific difficulties do arise, but in many cases they can be overcome.

Historically the need to introduce the concept of a multivalued function has been recognized by many mathematicians. Thus Euler, in his *Introductio in analysin infinitorum*, having defined a multivalued function as one "which receives several definite values when any desired value is substituted for the argument" [1, p. 7], then devotes several pages of his monograph to studying such a function, and as a special proposition states the following theorem: *If y is any function of z, then conversely z is a function of y* (p. 10). Thus Euler treats the inversion problem in the most general form, though for a restricted class of functions.

While Euler in [1] encountered multivalued functions in a general treatment of the inversion problem, Condorcet includes multivaluedness in the initial definition of the concept of a function. His statement was given in Sec. 6 above, and there is no need to reproduce it here.

It is likely that one could adduce many such examples. Multivalued functions acquired particular significance in the theory of functions of a complex variable, and the requirement of multivaluedness is usually included in the initial definition in textbooks on that subject (cf., for example, Lavrent'ev and Shabat [1, p. 17]). We limit ourselves to just a few words on the Weierstrass theory.

Weierstrass' point of departure was a power series with a positive radius of convergence (an element of an analytic function). Then carrying out all possible analytic continuations, he introduces the concept of an analytic function as the totality of series so obtained. Its domain of definition is the set of points at which these continuations converge, and its range is the set of all of their values. The values of the argument and the function together

2.13 Some other functional correspondences

formed what Weierstrass called an analytic structure (Gebilde).

What is important for us at this point is that under different continuations it can happen—and does, as a rule—that a given value of the argument yields not one but several, possibly infinitely many different values of the function. Thus for Weierstrass an analytic function is in general multivalued from the outset and becomes single-valued only under special assumptions; multivaluedness does not deprive it of mathematical determinacy, as Weierstrass himself specifically points out [1, p. 210].

Attempts have also been made to study multivalued functions of a real variable. Of these we note only the dissertation of Vasiliescu in 1925. Not having it at our disposal, we limit ourselves to a reference to the article of Marcus [1, pp. 1128–1129], noting that it attracted the attention of Lebesgue [20, p. 21, footnote]. Such functions arise in many investigations in analysis and the theory of functions; as an example we refer only to Lebesgue, who specifically defined an element-to-set correspondence: "We shall say that we have such a function [a multivalued function—F. M.] if to each value λ taken from a certain set where the function is defined some set of numbers is made to correspond; each of these numbers is represented by the symbol $\psi(\lambda)$" [20, p. 21].

It may be of interest to note that in very many arguments given in the theory of functions an element-to-set correspondence is used as an important fact in the course of the reasoning.

Of the works in which there is a general approach to the study of this type of functional dependence, we point out the following two. In a popular article [1] of 1935 Keyser discusses the identity of the three terms *relation*, *transformation*, and *function* as a self-evident thing, interpreting them in the most general form, i.e., he identifies, for example, a multivalued binary relation with a function. In 1966 Menger [1] gives an abstract exposition of the Weierstrassian analytic structure mentioned above.

Thus dependence of the element-to-set type, though not as widely used as element-to-element and set-to-element dependence, nevertheless has been and continues to be the subject of numerous studies. Whether such correspondences should be called functions or not is a secondary question; the kinship of these correspondences with the classical elementwise correspondences is not in doubt.

It may be that purists are even more suspicious of including in the class of functional dependences the fourth type of correspondence mentioned above, in which sets of one class are assigned to sets of another. But as in the preceding case, one can exhibit a large number of works in which such correspondences are studied and even name works containing formal definitions of them.

Transformations of set-to-set type have long been studied in mathematics, at least since the time when transformations of geometric figures were considered, for example projective transformations, but they received a particularly wide distribution in the fields of set theory and topology. The transition from a set to its complement, to its closure, to its interior, the mapping of one topological space into another, and the like—all these are examples of such correspondences. The theory of operations on sets also fits into this scheme. But evidently such correspondences are most prominently represented in such a large area of the theory of functions as integration of multivalued set functions, where the values of the set functions under consideration are numerical sets of a nearly arbitrary nature.[74]

As an example of works in which the two last types of correspondence (element-to-set and set-to-set) are studied in an extremely abstract form, we cite the article of Kuratowski "Les fonctions semi-continues dans l'espace des ensembles fermés" [1].

As for formal definitions of a functional correspondence of set-to-set type, we mention the definition of Carathéodory in 1918:

> The ultimate functional concept, which subsumes all the forms [of functional dependence—F. M.] studied up to now in analysis, is the following.
>
> To each point set A of some set $\mathfrak{A}$ of point sets there corresponds not a number but again some set B of the same space or a different one.
>
> To the collection $\mathfrak{A}$ of point sets under consideration there corresponds a collection $\mathfrak{B}$ of sets B, and we say that $\mathfrak{A}$ is mapped in a single-valued way onto $\mathfrak{B}$ [1, p. 73].[76]

In speaking earlier of a function as an elementwise correspondence, we mentioned several logical difficulties that arise in an attempt to give a formal definition. Evidently under a more careful study of set-to-element, element-to-set, and set-to-set correspondences the corresponding difficulties increase even more.

I do not know of any works that give a logical analysis of the definitions under consideration, and I suspect that such an analysis is impossible, at least at the present stage of development of logic. I also know of no works in which the relations between these four types of correspondences are explored;

---

[74] One must mention the curious investigations of R. Young in the 1920's and 1930's on the so-called "algebra of multivalued quantities," and multivalued integration. Of several papers by her on this topic we mention only [1].

[76] The words *point set* or *set of points* need not be understood in the sense that only sets formed by the points of a Euclidean space are considered.

2.13 Some other functional correspondences

one cannot agree with the assertion of Carathéodory that the definition he stated encompasses all forms of functional correspondence studied in analysis, although at first glance this does indeed seem to be the case: when the sets of his collections $\mathfrak{A}$ and $\mathfrak{B}$ are one-element sets, we clearly have an elementwise correspondence or an ordinary function; when $\mathfrak{A}$ consists of sets and $\mathfrak{B}$ of one-element sets, we have a set-to-element correspondence, or a set function; in the opposite case we have an element-to-set correspondence, i.e, a multivalued function; finally, when $\mathfrak{A}$ and $\mathfrak{B}$ both consist of sets, then we have multivalued set functions.

By its visualizability and simplicity such a picture apparently only masks the real difficulties; the requirements of the theory of types mentioned earlier are violated here even more explicitly. And if we add to that the fact that correspondences of mixed type are considered in mathematics, for example, when to a fixed set of a given family of sets and to a fixed element of a given set not necessarily belonging to the given family we assign an element of some other set, the situation becomes much more complicated.

It becomes still further complicated if we keep in mind that the history of the concept of a function is far from being as meager as we have represented it.

Even allowing for the basic limitation here to the framework of analysis and the theory of functions and further restriction that the modifications of this concept in the form of functionals and operators, distributions and hyperfunctions, and the various types of functions of the constructive theory, etc., are not being traced, the skeleton of not very sharply defined formal definitions we have marked out reflects very incompletely the full-blooded life of these definitions, even within the delineated framework. In all their modifications they have developed in close connection with unimaginably varied applications in various fields of mathematics and mathematical science, and with a general appreciation of the mutual relation of mathematics and other sciences. For example, the problem of the connection of the concept of a function with scientific law, which mathematicians, philosophers, and historians of science have pointed out many times, could by itself serve as the topic of an interesting separate work.

In the preceding sections of this essay we have given most of our attention to the conceptual aspect of the object of study. But no less interesting is its formulary, analytical aspect, which we intend to study, though only in part, in the remaining essays of this book.

Chapter 3
Sequences of functions.
Various kinds of convergence

3.1 The analytic representation of a function

In Sec. 3 of the preceding essay we saw that the concept of a function was distinguished as a separate object of study after methods were developed for representing it in the form of an analytic expression, as a formula containing a generally infinite number of rather simple, elementary mathematical symbols.

For a long time the only such elementary symbols were the powers of the variable, a sum of which defined the function. Then trigonometric functions were added to them, followed by various kinds of orthogonal polynomials and finally by general functions of some function class.

In its conceptual aspect this approach has roots deep in antiquity. It is akin to the Democritean and Archimedean decompositions of the object being studied into certain component parts of a simpler nature, whose study made it possible to draw conclusions about the object as a whole. Similarly the properties of a function represented by a series or sequence are obtained from the properties of the elements that compose it.

The analytic representation of a function is very convenient in that a convenient notation is usually available for the elementary symbols of which the function is composed and simple and often quite visualizable, transparent formal rules have been established, making it possible to carry out mathematical operations on them almost automatically. It sometimes suffices merely to calculate the sum of a series or the limit of a sequence of these elementary components in order to obtain a desired result about the function in question. It is this feature that makes the notion of analytic representation of functions so attractive to mathematicians.

The enrichment of the notion of a function traced above resulted in the necessity of dealing with at least four basic problems at each successive step. First, with each new enlargement of the class of functional dependences studied the question arose whether the stock of elementary functions that were serving for analytic representation of the functions of the class under consideration was sufficient, and how convenient the various elementary functions were. Second the need arose to find a method of analytically expanding a function being studied into elementary functions, a method of determining the coefficients of the series. Third, especially notable for the theory of functions, it was necessary to think about the meaning of the expression, "a function can be represented by a series or a sequence." Fourth and finally, once the meaning of a representation had been established, one needed to know whether that method of representation was sufficient, and also whether it was necessary in order to establish the desired properties of the function.

The first two of these problems were worked on together with the creation of analysis and at the same time helped to promote its creation. The introduction of formal power series into common use among mathematicians in the seventeenth century along with various methods for obtaining them—from simple infinite division to termwise differentiation and integration—kept this theoretical aspect of these foundational questions off the agenda for a long time. Every function was considered to be expandable in such a series and whether an investigator would be able to obtain the required representation seemed to depend only on his skill.

The debate over the vibrating string helped to make possible the inclusion of expansions in series of the functions $\sin nx$ and $\cos nx$ into the ranks of the simple elements. Such expansions came into an ever wider use and became the dominant ones in the nineteenth century. It was mainly with them that the problem of the coefficients, which has not yet been solved, turned out to be connected.

Other elementary functions began to be introduced in addition to the trigonometric functions: the polynomials of Legendre, Laguerre, Chebyshev, Bernshtein, the functions of Bessel, Sturm-Liouville, Haar, Rademacher, etc. Besides representations by more or less individual functions of such a type questions of the analytic representation of functions of some class characterized by a definite structural property, for example continuity, were studied.

No matter how great the variety of functions by which mathematicians were attempting to represent functions of a quite general type, however, progress was made in encompassing functions of ever larger classes only when the direction was changed and mathematicians began to analyze the nature of representation itself.

In the early days such an analysis had not especially worried mathematicians. As a rule the expansions had been formal and subject to formal operations. The question of clarifying the content of the concept of a representation did not become acute until a significant stock of contradictions had been accumulated, consisting generally of the loss of the former ease with which conclusions about the behavior of the function represented could be deduced from the behavior of the functions representing them in many cases of interest to mathematicians. But by the beginning of the nineteenth century

84   Chapter 3   Sequences of functions. Various kinds of convergence

there was a pressing need for such a clarification.

The first reaction to the need to answer the question, "What does it mean to say that a given series or sequence is an analytic representation of the function in question?" was the introduction of the concept of convergence at every point. This concept turned out to be too broad on the one hand, since it was very soon discovered that many properties of the representing functions do not carry over to the function represented under such a representation. For example, the limit of a sequence of continuous functions turned out to be discontinuous in general. Hence arose the problem of restricting the general concept of convergence at every point; and depending on specific needs mathematicians introduced more restricted kinds of convergence, such as uniform convergence, quasi-uniform convergence, absolute convergence, etc. On the other hand this concept fairly soon became insufficient, since not all the functions of interest to mathematicians could be defined analytically in this sense. In the enlargement of the concept of representation of a function by sequences of functions of a simpler nature three routes were marked out.

The first of these—not chronologically, but in conceptual simplicity—consisted of successive iterations of limiting passages at every point. In the early days such significant results were obtained on this road that mathematicians began to think that the only functions about which it was possible to speak of an analytic representation were functions obtained from continuous functions by such iterations, provided a transfinite set of iterations was allowed.

The second route consisted of broadening the concept of representability by giving up the requirement of convergence of the sequence at each point to the function being studied. A function came to be regarded as analytically represented if the sequence representing it converges to it everywhere except possibly on a certain set of points. In this way there arose the concepts of convergence almost everywhere, convergence in mean, convergence in measure, etc. Similarly not only convergence everywhere, but also its more restricted variants—uniform, absolute, and monotone convergence, etc.—were broadened. These specializations were sometimes combined with iterated passages to the limit.

Finally the third direction was the generalization of the concept of the sum of a series or the limit of a sequence by the introduction of various summation methods: Euler summation, Riemann summation, Cesàro summation, and others. A function is considered to be analytically represented by a given series if the series is summable to the values of the function by some summation method. Here, as in the second route, mathematicians either require the series to sum to the values of the function at every point or settle, for example, for summability almost everywhere, i.e., neglecting a set of measure zero. Again, as in the preceding section, summability is frequently restricted to be uniform or absolute, for example.

It should not be thought that the description just given reflects the historical process. The actual picture of the evolution of these ideas is much more complicated and interesting. Even a very large book would not suffice for a reasonably complete description of it. Thus the survey of Zeller [1] devoted entirely to summation methods lists about two thousand papers; the number of such papers has increased noticeably since Zeller's survey appeared, and the survey was not even a complete one. The number of works devoted to the various kinds of convergence is evidently of the same order of magnitude, and the number of studies on iterated limiting passages is perhaps only slightly smaller.

But the problem of the analytic representation of functions consists of more than just questions of representation by series or sequences. Other approaches to this problem have been and continue to be developed. Functions are representable by continuous fractions, integrals with a parameter, singular integrals, etc. Many investigators have worked on this and continue to do so.

The following pages contain a sketch of only particular elements of the colorful picture of the actual course of mathematical thought in the question of the analytic representation of functions.

Before passing to specific historical facts, one terminological remark will be helpful. The reader has probably noticed that the words *series* and *sequence* were used as synonyms above. In reality they are synonyms, since every series can be transformed into the sequence of its partial sums and conversely the terms of every sequence can be regarded as the partial sums of a series. The historical state of affairs is that the language of series was predominant earlier and the language of sequences later came to be used more and more often. We shall use the one or the other in what follows according to the work being discussed.

3.2 Simple uniform convergence

The history of the introduction of uniform convergence into analysis has been studied by many authors, among whom we mention Pringsheim [2, pp. 34–35], Hardy [5], Rosenthal [1, pp. 1137–1146], and Grattan-Guinness [2, pp. 112–123]. Sections 2–4 of this essay will be the longest sections in this chapter, yet even they will be far from exhaustive as a study of this history, since a voluminous literature on it already exists, some of which is accessible only with difficulty. In addition the reasoning of the mathematicians who worked with this concept is sometimes not very clear, and it is not possible to give it a single interpretation.

As stated, one of the principal devices, if not the principal device, for studying the objects used in analysis and the theory of functions is representation as an infinite sequence of simpler objects that are in some respect known and approach the object of interest more and more closely. The choice of the approximations can be carried out in very many ways, and the most valuable is a method by which the approximating functions convey the largest amount of information about the properties of the object being approximated. In particular, in many cases it is important to know when particular general properties of the approximating objects transfer to the object being approximated. Evidently the most convenient in this aspect, unsurpassed in a large variety of questions, are the uniform convergences of sequences of functions.

This last sentence is not a misstatement. The fact is that there are many forms of uniform convergence, not just one, as frequently seems to be the case in courses of analysis. For series converging everywhere on $[a, b]$ we shall follow Hardy [5, pp. 150–152] in distinguishing temporarily the following forms of uniform convergence.

Suppose given a series

$$\sum_{n=1}^{\infty} u_n(x), \qquad (1)$$

converging to a function $U(x)$ at each point $x \in [a, b]$; let $s_n(x)$ be its partial sums and $r_n(x)$ the successive remainders. The basic inequality among all the definitions we are considering is an inequality of the form

$$|r_n(x)| \leq \varepsilon. \qquad (A)$$

The first three definitions are the following:

A_1. *Uniform convergence on an interval.* The series (1) is said to converge uniformly on the interval $[a, b]$ if for every $\varepsilon > 0$ there exists $n_0(\varepsilon)$ such that inequality (A) holds for all $x \in [a, b]$ when $n > n_0(\varepsilon)$.

This is one of the most important and widely used forms of uniform convergence, and is the one to which the majority of textbooks on analysis are limited.

A_2. *Uniform convergence in a neighborhood of a point.* The series (1) is said to converge uniformly in a neighborhood of a point $\xi \in [a, b]$ if there exists a positive number $\delta(\varepsilon)$ depending on the point ξ such that on the interval $[\xi - \delta(\xi), \xi + \delta(\xi)]$ it converges uniformly in the sense of A_1. If $\xi = a$ or $\xi = b$, the interval $[\xi - \delta(\xi), \xi + \delta(\xi)]$ is replaced by a suitable one-sided neighborhood.

It is clear that if a series converges uniformly in the sense of A_1, then it converges uniformly in the sense of A_2 in a neighborhood of each point of

$[a, b]$. The converse of this statement is a nontrivial mathematical proposition that is proved by applying the Heine-Borel Theorem on finite coverings. The distinction between these two kinds of convergence is particularly important on the historical level, since, first of all, at the time of its introduction mathematicians were not yet using reasoning based on the Heine-Borel Theorem, and second, such reasoning is not always applicable.

A_3. *Uniform convergence at a point.* The series (1) is said to converge uniformly at the point $x = \xi$ if for every $\varepsilon > 0$ there exists a positive number $\delta = \delta(\xi, \varepsilon) > 0$ and a natural number $n_0 = n_0(\xi, \varepsilon)$ such that inequality (A) holds for $n > n_0(\xi, \varepsilon)$ and $x \in [\xi - \delta(\xi, \varepsilon), \xi + \delta(\xi, \varepsilon)]$.

The implications $A_1 \to A_2 \to A_3$ are obvious. However uniform convergence at a point does not in general imply uniform convergence in a neighborhood of the point. And although there is a theorem that states that uniform convergence of a sequence of functions at each point of an interval implies uniform convergence on that interval, its proof also relies on the Heine-Borel Theorem, and the argument involves an appeal to Zermelo's Axiom (Hardy, [5, pp. 151–152]).

In the present section we shall limit ourselves to the convergences A_1, A_2, and A_3.

The history of uniform convergence is usually dated from the year 1821, when Cauchy, in his *Analyse algébrique* [1, pp. 131–132] published a proof of the erroneous assertion that a series of continuous functions that converges everywhere has a continuous sum. In 1826 Abel [1, pp. 224–225, footnote] remarked cautiously that "this theorem seems to admit exceptions" and gave the example of the series[1] $\sum_{m=1}^{\infty} (-1)^{m-1} \frac{\sin mx}{m}$, whose sum is discontinuous at the points $x = (2k+1)\pi$. Somewhat earlier, in proving the theorem that

---

[1] It is of some interest to note that if Cauchy and Fourier had been more attentive to each other's work, Cauchy would probably not have made this error and Fourier would have had an opportunity to note it before Abel. Fourier had at his disposition examples of series of continuous functions that converge to functions that are discontinuous at individual points; in particular, the series exhibited by Abel is contained in his *Théorie analytique de chaleur* [1, p. 202], which also contains other expansions that are similar in this respect (pp. 219–220). Although this work appeared in 1822, it was reported piecemeal to the French Academy starting in 1807; and it is also curious that the French mathematicians who heard and read it failed to notice the error of their compatriot Cauchy, leaving that to the Norwegian Abel. One of the possible reasons for Fourier's oversight could be his idiosyncratic understanding of functional dependence: he thought that at points of discontinuity of the first kind of the functions he was considering the graph of the curve being represented contained vertical line segments erected at these points, i.e., he considered the function to be multivalued (cf., for example, [1, pp. 157–158]). With such an approach the curve

if a power series converges at the point $x = a$, it also converges for all x with $|x| < |a|$, Abel actually noted the property of a power series denoted by A_1 above, but not as a particular property of the series, rather as a conclusion following from another proposition of his known in modern literature as Abel's Lemma [1, p. 223].

In 1841–1842 Weierstrass [3, 4] used this property of power series and even said that the series converges uniformly (gleichmässig oder gleichförmig convergiert). However in [3] it is not inequality (A) that is placed at the foundation of the definition of uniform convergence of a series, but rather the other way round: inequality (A) for all the values of the argument in question is deduced from the assumption of uniform convergence, which is not defined at all [3, pp. 68–69]. Most likely Weierstrass applied the term *uniform convergence* not to convergence in the sense of A_1, but to the property of power series expressed by the theorem of Abel given above, since the convergence of a power series in the sense of A_1 does indeed follow from that theorem. Only in a modern reading does his theorem on series of analytic functions [3, pp. 73–74] look like a theorem on absolutely and uniformly convergent series. There is also no definition of uniform convergence in the second of Weierstrass' works [4], dating from 1842. This meaning can be gleaned only from complicated reasoning, and moreover it appears there only once [4, p. 81] and in such close connection with absolute convergence that one gets the impression that it follows from the latter. Thus it is unlikely that Weierstrass had any very clear idea of uniform convergence when he wrote [3] and [4], and so the connection of these works with the concept of uniform convergence of series is very hypothetical and relates mostly to terminology that still had a very narrow meaning at the time. Moreover these papers were not published until 1894 and could not have had any influence on other mathematicians.

The first works with which the concept of uniform convergence is usually linked are the articles of Seidel [1] and Stokes [1].[2] However, the frequently encountered assertion that these mathematicians introduced the actual concept of uniform convergence is wrong,[3] since first of all no single concept of

---

representing the function $\sum_{m=1}^{\infty}(-1)^{m-1}\frac{\sin mx}{m}$ does not look at all discontinuous. Fourier was not alone in this understanding of the behavior of the sum of a series at a point of discontinuity; some mathematicians after him also thought this way, for example, Duhamel and Du Bois-Reymond. Cf. Paplauskas [1, p. 162].

[2] In regard to the dates of these works opinions differ. Some say 1847, others 1848; some assert that Stokes' work appeared first, other say Seidel's. We are not in possession of the original publications.

[3] Cf., for example, Wieleitner [1, p. 385].

uniform convergence exists and second, even in relation to some particular type of uniform convergence it is impossible to say categorically that it was consciously introduced by them. Rather, both of them established only the fact of nonuniform convergence of a series of continuous functions in neighborhoods of the points at which the sum of the series has a jump discontinuity. To be sure, we can now read into their reasoning the idea of some form of uniform convergence, but nearly the same can be said about Cauchy's reasoning in 1821.

We shall now pause to discuss the paper of Seidel [1].[4]

Seidel's point of departure is the comparison of Cauchy's 1821 theorem with the results of Dirichlet on Fourier series representation of discontinuous functions, and his main result is stated in the form of a theorem: *Given a convergent series representing a discontinuous function of the variable x whose terms are continuous functions, in the immediate neighborhood of a point at which the function has a saltus, there is a value of x at which the series converges arbitrary slowly* [1, p. 37].

It follows directly from the statement that one should not look for uniform convergence on a whole interval (convergence A_1) in Seidel's writing. He is interested in the behavior of the function in a neighborhood of a point of discontinuity. If we interpret Seidel's judgment in the language adopted above (his own statements are lengthy and not very clear), he takes arbitrarily slow convergence to mean the following:

Let ξ be the point in question and $\varepsilon > 0$ a given number. We take some neighborhood $[\xi - \delta, \xi + \delta]$ of the point ξ. Since the series (1) converges at all points of this neighborhood, for every $x \in [\xi - \delta, \xi + \delta]$ there exists a natural number $n_0(\varepsilon, x)$ depending on the ε chosen and the x in question, such that the inequality holds for every natural number $n > n_0(\varepsilon, x)$. Seidel distinguishes two possibilities: either among these numbers $n_0(\varepsilon, x)$ corresponding to the different points of the neighborhood there is a largest N, and such an N can be found for each $\varepsilon > 0$, or there is no largest N among these $n_0(\varepsilon, x)$, and no such can be found, no matter how much we shrink the neighborhood of the point ξ. This last case is what Seidel calls arbitrarily slow convergence at the point ξ [1, p. 41].

The proof of this theorem of Seidel's is interesting in that it shows correctly from the beginning [1, p. 48] that if the first possibility holds, then the series of continuous functions does converge to a function that is continuous in a neighborhood of the point in question. In other words it is shown here that convergence A_2 of a series of continuous functions is sufficient for its sum to be continuous in a neighborhood of the point. The very lengthy arguments

---

[4] Stokes' paper will be discussed in the following section.

of the second part reduce to the following: since there exist series of continuous functions that converge to discontinuous functions, such series cannot converge in the sense of A_2 in a neighborhood of a point of discontinuity, since otherwise, by what was proved in the first part, they would converge to continuous functions in a neighborhood of that point.

Thus, although Seidel was not stating specifically the concept of convergence A_2, he nevertheless essentially defined it in the course of the reasoning in the proof of his theorem and gave it an important application.

We note further that Seidel did not regard arbitrarily slow convergence of a series of continuous functions to be a criterion for discontinuity of the sum of the series: "without further investigation one cannot exclude the possibility that this same situation occurs with series whose values do not undergo a jump" [1, p. 44]. Subsequently such a situation was actually discovered.

Seidel's article went unnoticed for quite some time. This is partially explained by the small circulation of the journal in which it was published (*Sitzungsberichte der Bayerischen Akademie der Wissenschaften*) and partly because of the obscurity of the author; the main reason, however, was evidently that he considered a type of uniform convergence, moreover in a form that was not very accessible, for which the need did not arise until much later.[5]

It is more surprising that Cauchy's 1853 memoir [4] also went unnoticed, although in this memoir, correcting his erroneous theorem of 1821, he explicitly introduced uniform convergence in the sense of A_1 and proved that when uniform convergence is present the sum of a convergent series of functions of either a real or a complex variable is a continuous function on an interval or a region. To be sure, instead of the inequality (A) Cauchy used the inequality $|s_{n'}(x) - s_n(x)| < \varepsilon$, $n' > n$, and merely remarked that when the latter holds, the inequality $|r_n(x)| < \varepsilon$ also holds, but it can be shown that this is the same thing as uniform convergence.[6] As Cauchy noted, the impetus for this paper was a remark of Briot and Bouquet that the statement of his 1821 theorem is correct only for power series. These last-named authors probably based their remark on the abovementioned proposition of Abel, since Cauchy's paper involves the same series of sines used as an example by Abel. Thus we can assert that it was Cauchy who first gave the definition A_1, without having any special name for the new concept.

However, as we have said, mathematicians somehow "overlooked" Cauchy's memoir [4], and a cycle of investigations began on how to introduce uniform convergence.

---

[5] Much the same can be said about Stokes' memoir.
[6] Cf., for example, Luzin [9, pp. 243–245].

3.2 Simple uniform convergence

Such an introduction is usually connected with the name of Weierstrass, and Pringsheim [2, p. 35], for example, says directly: "The name *uniform convergence* is due to Weierstrass, and it is because of his lectures that the crucial significance and necessity of this concept became generally known."

It is not known when Weierstrass began to use uniform convergence in his lectures. He began to lecture at the University of Berlin in 1857,[7] i.e., after the publication of the memoirs of Seidel [1] and Cauchy [4], and one may surmise that they had become known to Weierstrass by that time since, for example, he had begun to take an interest in Cauchy's investigations in 1842. Despite this assumption, the role of Weierstrass in the creation of the concept of uniform convergence must evidently be considered very significant. It seems to us that the situation was most likely the following: after becoming familiar with Cauchy's memoir [4], Weierstrass simply extended the useful term he had adopted earlier to the more general type of convergence proposed by Cauchy. But he did something much greater in his lectures.

Seidel, Stokes, and Cauchy approached the study of uniform convergence only in connection with the question of the continuity of the sum of a series of continuous functions, and none of them connected this concept with other problems of analysis. Moreover in 1823 Cauchy (and Stokes after him) stated another erroneous theorem, that a convergent series of continuous functions can be integrated termwise (Cauchy [2, pp. 221–222]).[8] Since it turned out to be much more difficult to detect the error in this case, it seems that no mathematician before Weierstrass doubted its correctness, and such series were integrated termwise, the operation being carried out as a self-evident thing. Weierstrass, in contrast, showed in this lectures that uniform convergence is important here also, and as a result this concept ceased to be dependent on the particular application it had previously been given.[9] The situation was similar in regard to termwise differentiation of series.

So far as is known at present the first mention in print of a connection between uniform convergence and termwise differentiation and integration of series occurs in an article of L. Thomé [1] in 1866. In studying the hypergeometric function of Gauss he needed to show the possibility of termwise differentiation of a series. To do this he states, though not very clearly, the definition of uniform convergence, and, noting that it is present for the series under consideration, continues, "But if we differentiate termwise a series that converges uniformly in a connected region and a convergent series is obtained, then the latter, as is known by the theorem on termwise definite integration

---

[7] Cf. Dugac [1, p. 54].
[8] Seidel does not discuss this question in [1].
[9] For yet another of Weierstrass' contributions to these questions see below.

of a uniformly convergent series, is the derivative of the function defined by the original series" [1, p. 324].

This proposition is interesting for at least two reasons. First, in it uniform convergence is simultaneously connected with three important problems of analysis—conditions for the continuity of a sum, and termwise differentiation and integration. Second, these connections are regarded as known and the author makes no attempt to justify them. If, as we saw above, one could say this with reference to continuity, nevertheless for differentiation and integration Thomé's words "as is known" most likely referred to Weierstrass' lectures. Indeed the author points out that he had attended these lectures and that it was Weierstrass who caused him to study the questions to which his work was devoted [1, p. 322, 324]. In regard to termwise integration Heine [1, p. 353] confirms this.

Heine [1] adopted the concept of uniform convergence as a tool in 1870. Noting the unsatisfactory foundation of the proposition on termwise integration of an everywhere convergent trigonometric series and casting doubt on practically the whole theory of trigonometric series, which was to a large degree based on this proposition, he made an attempt to put a firmer foundation under it, starting with the theorem on uniqueness of representation by trigonometric series. It was for this reason that he first of all introduced the concept of uniform convergence. Referring the reader to the book of Paplauskas on this matter [1, pp. 217–219], we shall make only a few comments.

Heine [1, p. 356] introduced the concept of uniform convergence in the form in which Cauchy had defined it. In this place he did not refer to anyone, although at the very beginning he noted that Weierstrass "has remarked that the proof of this proposition [on termwise integration—F. M.] requires not only that the series converge within the limits of integration, but that it converge uniformly" [1, p. 353]. In a footnote on the same page he mentioned the work of Thomé [1], and in a footnote on p. 355 the article of Seidel [1], known to him from his reading of Cantor. It is of some interest that, although Thomé understood uniform convergence rather in the sense of A_1 with an explicit estimate of the remainder of the series, while Seidel had in mind convergence in the sense of A_2, Heine, though referring to their work, chose Cauchy's form as the basic definition, i.e., A_1 with the estimate $|s_{n+m}(x) - s_n(x)| < \varepsilon$, but didn't mention Cauchy.

One of the interesting aspects of Heine's work is the conscious emphasis on the principle of neglecting point sets in studying the various questions of analysis. To be sure, he neglected only finite sets, but he understood the principle itself in a very broad sense. Thus, after the definition of uniform convergence he introduces the concept of "a generally uniformly convergent series," i.e., a series that converges uniformly on $[\alpha, \beta]$ if arbitrarily small neighborhoods of a finite number of points are removed from $[\alpha, \beta]$ [1, p. 356]; he speaks of a "generally continuous function" [1, p. 355]; he formulates and proves his three main theorems in accord with the same principle; the behavior of the series at critical points does not interest him [1, p. 356].

Like Seidel, Heine leaves open the question whether a series of continuous functions that converges to a continuous function at every point must converge uniformly to it [1, pp. 353–354].[10] A negative answer was given in 1874 by Du Bois-Reymond [2, p. 43, footnote]. Nevertheless in the same year Stolz attempted to prove that uniform convergence of a series of continuous functions is necessary in order for the sum to be continuous. We are not in possession of his work[11] and we do not know what kind of uniform convergence he had in mind and whether he was attempting to prove continuity of the sum at a point or on an interval. Cantor's criticism [6] of his proposition in 1880, involving the construction of a counterexample, was too late, since Darboux had already constructed such an example in 1875, which will be discussed below.

If Heine, having adopted uniform convergence as a tool for studying the theory of trigonometric series, proved in particular a uniqueness theorem for the case of generally uniform convergence, Cantor [1] in that same year of 1870 stated and attempted to prove a more general uniqueness theorem, abandoning the hypothesis of uniform convergence. However, he did not succeed in doing so without the hypothesis of uniform convergence in [1], and it was only in the following year that, using a remark of Kronecker, he first outlined [2] a more correct proof, and then carried it out in the required generality [3].

Since he was applying uniform convergence in [1], he defined it [1, p. 80], and in a form slightly different from that of Heine. To be specific, he used the form A_1 with an estimate of the remainder, and this was evidently the first publication of the formal definition later accepted by the majority of mathematicians. In this connection it is of interest that although Cantor knew of Seidel's work and, as noted, Heine even referred to it, Cantor, like Heine, preferred convergence A_1 over A_2. From Cantor's reasoning it is particularly apparent that the decisive factor in this choice was that he needed uniform

---

[10] We remark in passing that in 1871 Du Bois-Reymond [1], evidently not yet knowing of these studies of uniform convergence but starting from just the remark of Abel on the incorrectness of Cauchy's theorem, established a sufficient condition for the continuity of the sum of a series of continuous functions not directly connected with uniform convergence. His condition was that the series of coefficients of the series converge absolutely, and the functions themselves—both the terms of the series and the sum—be bounded.

[11] Published in *Berichte des naturwissenschaftlichen und medizinischen Vereins in Innsbrück*, December 1874, and known to us from the work of Cantor [6].

convergence to integrate a series termwise on an interval [1, p. 83]. Thus two separate questions of analysis—continuity of the sum of a series of continuous functions and termwise integration—led to the introduction of two different forms of uniform convergence. Since the second question was the subject of more acrimonious discussion in the ensuing years, especially in the stormy development of the theory of trigonometric series, it is not surprising that convergence A_1 became the more important.

Besides all this Cantor's papers [1–3] probably weakened to some extent the interest of mathematicians in uniform convergence, since their aim was to get rid of it in the important question of the theory of trigonometric series.

As it happened the turning point in the relation of mathematicians to uniform convergence occurred in the years 1874–1875. In the 1874 paper mentioned above Du Bois-Reymond, developing Heine's doubts regarding the foundations of the theory of trigonometric series, wrote:

> For that reason the theory of trigonometric series, in which termwise integration plays an important role... has been struck in its very heart by this new concept [uniform convergence—F. M.] and moreover in such a way that as a result of this blow we—in relation to the most important theorems of this theory—return not only to the point of view of Dirichlet, but even to the point of view of Fourier [3, p. 45].

He places the concept of uniform convergence at the basis of his exposition and besides this example uses it in studying a variety of questions of the theory of trigonometric series. In 1875 the concept acquires a more general meaning. This refers particularly to Darboux' classic "Mémoire sur les fonctions discontinues" [1].

Darboux knew the works of Thomé, Heine, and Cantor. As the definition of uniform convergence he took A_1, emphasizing in doing so that it applies equally to everywhere-convergent series of both continuous and discontinuous functions [1, p. 77]. He proves once again more distinctly that a uniformly convergent series of continuous functions has a continuous function as its sum [1, pp. 77–78] and shows by the example

$$\sum_{n=1}^{\infty} [n^2 x^2 e^{-n^2 x^2} - (n+1)^2 x^2 e^{-(n+1)^2 x^2}] = x^2 e^{-x^2}$$

that the condition of uniform convergence is not necessary for the continuity of the sum. In addition Darboux proved that a uniformly convergent series of functions having one-sided limits converges to a function of the same type and illustrated this with several examples [1, pp. 79–82].

He then proved that a uniformly convergent series of Riemann integrable functions can be integrated termwise, i.e., the termwise integrated series converges and moreover to the integral of the sum of the series [1, p. 82–83]. By the example of the series

$$\sum_{n=1}^{\infty}[-2n^2 xe^{-n^2 x^2} + 2(n+1)^2 xe^{-(n+1)^2 x^2}] = -2xe^{-x^2}$$

Darboux establishes that if the condition of uniform convergence is not met, then in general it is not possible to integrate termwise even a series of continuous functions converging everywhere to a continuous function, thereby overturning Cauchy's 1823 theorem; in this example the integral of the sum naturally exists, but is not equal to the sum of the integrals of the terms.

He connected uniform convergence with termwise differentiation also. For this, however, uniform convergence of the original series turned out to be insufficient; rather it was necessary that the termwise differentiated series converge uniformly and that its terms be integrable.

In the same year of 1875, independently of Darboux, Thomae[12] proved a theorem on termwise integration [1, pp. 47–48], also using the Cantorian form A_1 of uniform convergence.

The publication in the 1870's of the works of Heine, Cantor, Du Bois-Reymond, Darboux, and Thomae, which contained studies and applications of the concept of uniform convergence in the form A_1, caused this concept to become one of the most important in analysis and the theory of functions and to occupy a place of honor in subsequent studies down to the present day.

In 1878 Dini made a fundamental contribution to the development and application of uniform convergence [4]. In his book we encounter not only the convergences A_1, A_2, and A_3, but also more general convergences that will be examined in the following section. For the time being we limit ourselves to the convergence A_2.

As was noted above, Seidel actually introduced convergence A_2, but did not distinguish it as a special kind of convergence. Dini also did not distinguish it explicitly, although he used it, and in particular proved that a series of functions having one-sided limits and uniformly convergent in a neighborhood of a point, converges in a neighborhood of the point to a function that also has one-sided limits. Moreover he has in essence here a kind of uniform convergence even more general than A_2, since he was considering only left- or right-neighborhoods.[13] He also went a step beyond Darboux in the ques-

---

[12] Not to be confused with the Thomé mentioned above.

[13] We remark in passing that in this theorem and in many others Dini considered not only functions defined on intervals but also functions whose domains of definition are unbounded sets.

tion of termwise differentiation, proving that if the series (1) converges in a neighborhood of a point ξ and each term has a bounded derivative and the differentiated series converges in the sense of A_2, then the original series can be differentiated termwise [4, pp. 115–116].

In a certain sense the ultimate step in relation to uniform convergence A_2 was taken by Weierstrass. In his paper [1] he introduced explicitly only convergence A_1 (p. 202, footnote), and did not state A_2, obviously considering the latter unnecessary. His grounds for doing so were quite weighty, since he proved (pp. 203–204) that if the series (1) converges uniformly in a neighborhood of each point situated inside or on the boundary of a simply-connected region, then it converges uniformly on all of this region. Weierstrass' proof deserves particular attention because of the importance of the argument used in it.

The reasoning proceeds as follows. Suppose given a planar region B, and let a and a' be any two points of this region, of which a is in a neighborhood of a' of radius R. Let $D = |a - a'|$ be the distance between the points a and a'. If the series in question converges uniformly on a neighborhood of each point inside or on the boundary of the region B, then the neighborhood of the point a' in which this series converges uniformly can be chosen so that its radius R' will satisfy the inequality

$$R - D \leq R' \leq R + D,$$

and R' will not vanish for any point inside or on the boundary of the region, since otherwise the series would not converge uniformly in a neighborhood of this point. Hence the radius of the neighborhood is a continuous positive-valued function of a variable point of this region. This function attains a minimum value $R_0 > 0$ at one point (at least) inside or on the boundary of the region:

> Therefore the region B can be partitioned into a finite number of parts of a form such that in each individual part the greatest distance between any two parts is less than R_0. Then each of these parts lies in a neighborhood of some arbitrarily chosen point of it; consequently for the values of x belonging to that part the series converges uniformly, from which it follows immediately by what has been said that the theorem is true [1, p. 204].

We would reason this out much more simply nowadays. Since the series converges uniformly in a neighborhood of each point inside on the boundary of the region B, the closed set $\overline{B}$ is covered by neighborhoods of uniform convergence. Then by the Heine-Borel theorem a finite set of neighborhoods of this covering system can be selected that still cover $\overline{B}$. It follows immediately

3.2 Simple uniform convergence

from this that the series in question converges uniformly in the entire region B.

But Borel [1] did not publish the finite-covering theorem until 1895, and moreover stated and proved it only for countable coverings; the proof for uncountable coverings was published by Young in 1902 and Lebesgue in 1904.[14] Weierstrass could not wait for the Heine-Borel theorem to appear, and he actually proved it in the reasoning just given, moreover not for a closed interval, but for a closed planar region and for uncountable coverings.[15] In 1882 Pincherle [1, p. 67] stated this theorem as a separate proposition and proved it by Weierstrass' method. To be sure, he originally stated it in the following form:

> If to each point x_0 of a region C and its boundary there corresponds one and only one value of a quantity X (a function of x in the most general sense of the term) and if one can determine a neighborhood of the point x_0 such that the upper bound of the absolute values corresponding to this neighborhood is a finite number $L_0(x_0)$, then there will exist a number N such that $|X| < N$ in the entire region.

The form of the statement was shaped by the questions that interested Pincherle in [1]: he was studying families of uniformly bounded functions. But he also noted here [1, pp. 67–68, footnote] that such a theorem holds when the lower bound of the values of X in a neighborhood of the point is studied. The theorem asserts that when this bound is different from zero, there exists a positive number M such that $|X| > M$ at all points of the region. He pointed out that this theorem can serve as a proof of the Weierstrass theorem mentioned above, and also as a proof of a theorem on uniform continuity.

Indeed, if the quantity X in Pincherle's last formulation is, for example, the length of an interval covering a given point of the closed interval, then it immediately implies the existence of a finite covering, i.e., the Heine-Borel theorem. For that reason Hildebrandt was too cautious when he wrote [1, p. 424] that Pincherle has the Heine-Borel theorem here only in embryo. On the contrary in Borel's result we find rather a mere corollary—an important one for applications, to be sure, but nevertheless a corollary—of Pincherle's theorem.

Since the present section has already gone on too long, we shall try to be briefer about the convergence A_3.

Here one should evidently begin with the interesting paper of Du Bois-Reymond [7]; this paper is so idiosyncratic, however, that it would require too

---

[14] Cf. Hildebrandt [1, p. 425].

[15] We are by no means claiming that Weierstrass was the first to apply such reasoning.

much space to discuss it. Moreover its author is not always clear, and it is not always possible to give a unique interpretation to his results and reasoning. It is indubitable, however, that along with a rather clumsy exposition of known facts, this paper contains many interesting ideas and merits separate study.

We limit ourselves to the comment that here Du Bois-Reymond introduced the convergence A_3 under the name of uniform convergence [7, p. 335]. It was used by Pringsheim [1, pp. 64–65, 80–81], and in the form in which it was defined at the beginning of this section it was introduced by W. H. Young [1, p. 90]. In the same paper Young studied one-sided convergence in the sense of A_3.

Starting from the definition A_3, Young introduced uniform convergence on a closed interval or a closed set as convergence A_3 at each point of such a closed interval or set. Since such a definition differed from the already established definition A_1, he proved that uniform convergence on a closed interval (or set), as introduced above, becomes A_1 on the closed interval (or set, if the statement of A_1 is suitably generalized). To do this he used a generalized Heine-Borel theorem that he had created in a previous paper[16] [1, p. 90]. Young's definition of A_3, besides being more general in the sense of being applicable to arbitrary closed sets, is also convenient because it extends literally to intervals and open sets, an extension he does carry out [1, p. 91]. Young also noted here that uniform convergence at a point in general does not imply uniform convergence in an arbitrarily small neighborhood of that point.

After Young convergence A_3 was studied by many mathematicians,[17] among whom F. Riesz, for example arrived at the concept in 1908, not knowing of the work of Pringsheim and W. H. Young, to which he referred only later (F. Riesz, [11, p. 214]). Some mathematicians, even as acute as Pringsheim, conflated the convergences A_2 and A_3, and Riesz was obliged once again to explain the difference between the two [11, p. 214].

The study of the convergences A_1, A_2, and A_3 in their various aspects—the distribution of the points of uniform and nonuniform convergence, the character of the set of these points, and the like—was carried on by many people: Osgood, W. H. Young, Hobson, Riesz, Hahn, and others.[18]

---

[16] Published in the *Proceedings of the London Mathematical Society*, **35** (1902), pp. 384–388; we have not had access to this work. Cf. Hildebrandt [1, p. 425].

[17] W. H. Young himself studied it, for example, in a paper published in the same *Proceedings* (series 2, **6**, 1906, pp. 29–51), to which we also have not had access.

[18] Voluminous, though nowadays far from complete, bibliographical references can be found in the book of Rosenthal [1, pp. 1140–1143].

3.3 Generalized uniform convergence

The three forms of uniform convergence considered above have in common the fact that the inequality

$$|r_n(x)| < \varepsilon \qquad (A)$$

is required to hold for *all* n, from some natural number on. If we weaken this requirement in the sense of settling for inequality (A) only for an infinite number of natural numbers n larger than some natural number n_0, but not necessarily all such integers, we obtain three more kinds of uniform convergence, which again, following Hardy [5, p. 152], we shall denote B_1, B_2, and B_3. The definitions of these convergences differ from the definitions A_1, A_2, and A_3 respectively only in that the weakened requirement is imposed on inequality (A) in the former.

The convergences B differ from the convergences A in one crucial point. As we have said, convergence A_3 at every point of a closed and bounded set implies A_1 on this set, and hence implies A_2 also. No such implications hold for the convergences B. The reason is that the conditions for uniform convergence at each point involve only the existence of some subsequence of the natural numbers $n_1 < n_2 < n_3 < \cdots$. But this in general does not imply that one can construct a single sequence applicable to all the points of the set under consideration, as required for B_1.

The history of the convergences B_1, B_2, and B_3 should evidently begin with the memoir of Stokes [1], since he was the first to arrive at one of them. The part of his memoir containing a discussion of uniform convergence was the subject of a special article of Hardy [5], and we shall therefore limit ourselves basically to a summary of the results of the latter.

We begin by noting that Stokes' exposition is far from clear and even Hardy was obliged to accompany his analysis with words like "if we have understood him correctly," "it seems to me," and the like. Therefore one can speak only of an interpretation of Stokes' reasoning.

According to Hardy's interpretation Stokes introduced the convergence B_2 and attempted to prove that this convergence is necessary and sufficient for the sum of a series of continuous functions converging everywhere in a neighborhood of a point to be continuous at the point in question. His proof of sufficiency is correct (and this must be recognized as a great achievement for the time), but he made an error in the proof of the necessity. Moreover Stokes did not connect the concept of uniform convergence with termwise integration of series, although he used the latter technique in his memoir and stated the corresponding erroneous theorem of Cauchy with a reference to Moigno.

To this one should add that the basic concept for Stokes, as for Seidel, was infinitely slow convergence, and convergence B_2 occurs in his work in the proof of his proposition [1, p. 281] without being explicitly distinguished.

While mathematicians began to refer to the paper of Seidel in 1870, Stokes' memoir remained unknown (in regard to uniform convergence) still longer. Besides the factors mentioned in the preceding section this is partly due to the fact that the type of convergence he introduced was not of particular interest at the time. Mathematicians were as yet more interested in convergence of type A, mainly A_1.

Nevertheless facts established in the 1870's, such as the nonnecessity of convergence A_1 for the continuity of the sum of a series of continuous functions or for the legitimacy of termwise differentiation and integration could not help stimulating attempts to weaken the overly rigid requirement of convergence A. In 1878 Dini [4] obtained some crucial results in this direction. Besides the convergences A all three kinds of convergence B are apparent in his book. Dini, however, gave a formal definition only of convergence B_1 [4, p. 103], and the others are contained in the statements of his theorems and in the reasoning given in their proofs. His definition is the following:

> We shall say that the series $\sum u_n$ converges *simply uniformly* on an interval (α, β) or for values of x that can be considered in this interval[19] if for each positive and arbitrarily small number σ and each number m' there exists an integer m larger than m' such that for all values of x that can be considered between α and β, including α and β, the remainder R_m corresponding to this m is numerically less than σ [4, p. 103].

This definition differs from B_1 in form, but if we take into account that m will also change when m' is chosen, we arrive at B_1. This concept was subsequently defined in both ways.

It is clear that B_1 is no less general a form of convergence that A_1, and Dini immediately points this out [4, p. 103]. But he did not know whether B_1 is really more general than A_1. He succeeded in proving that for series with positive terms these convergences coincide [4, p. 103], but he doubted whether they were fully equivalent. For that reason he stated and proved his theorems separately for the various kinds of uniform convergence.

Dini established many theorems about the convergences B. The most interesting of these seems to be his Theorem III: *If the terms of a series $\sum u_n$ are functions of x on a given interval (α, β) or depend on a variable x that can assume an infinite number of values for which a is a limit point, and if*

---

[19] That is, if the functions are defined on some set contained in (α, β).

the limits u'_n of these terms u_n from the right or left for $x = a$, for example from the right, are defined and finite, then in order for the limit of the sum U^x of the series $\sum u_n$ at $x = a + 0$ to be defined and finite, it is necessary and sufficient that: 1) the series of limits $\sum u'_n$ be convergent; 2) for every arbitrarily small positive number σ and every arbitrarily large integer m' there exist two numbers ε and m, for which the first is nonzero and positive and the second is an integer larger than m' such that for all values of x that can be considered between a and $a + \varepsilon$ (including a) the remainder $R_m^{(x)}$ of the series $\sum u_n$ is numerically less than σ [4, pp. 107–108].

This theorem is interesting first of all because its statement contains the explicit definition of uniform convergence B_3, or more precisely the definition of one-sided convergence of this type, and second because in the special case that occurs when the terms of the original series are continuous functions in a neighborhood of the point a, i.e., the limits from the right and left of the values of these functions coincide, convergence B_3 turns out to be a necessary and sufficient condition for continuity of the sum of the series at the point; this is the theorem that Stokes, whose work Dini seems not to have known, was trying unsuccessfully to prove. It is also curious that Theorem III was of interest to later mathematicians only in the case of continuous functions, not in its full generality; and, so far as I know, only Hobson [3, pp. 128–129] has paid any attention to this theorem.

It would seem natural that after Theorem III in its continuous variant one could also state the following proposition: a necessary and sufficient condition for the continuity of the sum of a series of continuous functions that converges on $[\alpha, \beta]$ is that the series converge B_3 at every point $x \in [\alpha, \beta]$. However, by the remark made at the beginning of this section no such inference can be made from Theorem III and the subtle analyst Dini, either having discovered this fact or having sensed it, concluded only [4, pp. 109–110] that convergence B_2 is sufficient for the sum of a series of functions continuous to be continuous in the neighborhood of a point, while convergence B_1 is sufficient for continuity of the sum on the entire interval.

Dini applied this convergence also in the problem of termwise differentiation [4, pp. 112–117]. The Dini conditions are very cumbersome and even 50 years later Hobson was unable to simplify them to any degree when he expounded them in his monograph [3, pp. 332–338]. For that reason we shall not go into detail about them. In studying termwise integration Dini limited himself to series convergent in the sense of A_1 [4, pp. 383–386] and remarked only that his proof of the theorem on the possibility of termwise integration of a series that is uniformly convergent in the A_1 sense implies that if the series in question converges in the B_1 sense, then its sum is integrable. He left open the question of conditions under which the sum of the integrals equals

Chapter 3 Sequences of functions. Various kinds of convergence

the integral of the sum.

After Dini many people took up the study and application of the B convergences. We shall examine only a few works.

As was stated above, Dini had been unable to answer the question of the relation between the convergences A_1 and B_1. In 1881 this question was answered by Volterra, who exhibited the series $\sum u_n(x)$ with terms

$$u_{2n-1}(x) = x^{n+1}, \quad u_{2n}(x) = -x^{n+1}\left\{1 - \frac{1}{(n+1)!}\right\},$$

for $0 \leq x < 1$ and $u_n(1) = \dfrac{1}{(n+1)!}$, which converges B_1 but not A_1 [1, p. 10] on the closed interval $[0,1]$. In this same paper he proved that a series of pointwise discontinuous functions that converges B_1 may converge to a totally discontinuous function.[20]

A significant contribution to the study of convergence B_1 was made by Bendixson [1]. Starting from the book of Dini, he first showed [1, pp. 605–607] that there exists a whole class of series that converge in the sense of B_1 but not uniformly in the sense of A_1 on an interval; the example of Volterra given above is a particular series of this class (Bendixson probably did not know of Volterra's work). He then constructed an example of a series of continuous functions that converge to a continuous function but not in the sense of B_1, thereby confirming Dini's general theorem that B_1 is only a sufficient condition for continuity of the sum [1, p. 608]. In regard to termwise integration Bendixson limited himself to the easily proved theorem that convergence B_1 of a series of continuous functions is sufficient to allow termwise integration when the series of integrals converges [1, p. 609]. He also proved a generalization to the case of B_1 convergence for the Darboux-Dini theorem on termwise differentiation of series [1, pp. 609–611].

In addition Bendixson established criteria for the convergences A_1 and B_1 of series [1, pp. 611–615] and applied them to simplify the proof of the existence of integrals of a system of ordinary differential equations [1, pp. 616–622]. Such criteria, especially for convergence A_1, were studied by many people, and we shall not dwell on them.[21]

Arzelà then studied the B convergences even more profoundly [5]. His memoir "Sulle serie di funzioni," is devoted mainly to another form of uniform convergence that we shall take up in the following section, but it contains

---

[20] The concept of a pointwise-discontinuous function as a function with an everywhere dense set of points of continuity was introduced in 1870 by Hankel [2]; he called functions totally-discontinuous whose points of discontinuity fill an entire interval of the region of definition.

[21] Some of them are studied in the book of Hobson [3, pp. 115–118].

some very interesting results on B convergence as well. Arzelà's exposition is carried out in a mathematical language that is different from the more widely used language of Hardy adopted in the present book and nearer to that in which Du Bois-Reymond expressed himself in [7]. For that reason, to shorten the exposition, we shall confine our remarks in this section and the next to the results of Arzelà that are comparatively simple to express in the terminology we have adopted and which in general are special cases of more general considerations of this author, as he himself usually pointed out.

First of all Arzelà reexamined from a more general point of view many fundamental results relating to A and B convergence. In particular in [5] he restated Dini's theorem on necessary and sufficient conditions for continuity at a point of the sum of a convergent series of continuous functions [5, pp. 142–145], analyzed numerous examples illustrating the relationship between the various kinds of uniform convergence [5, pp. 153–171], reinterpreted the results of Bendixson [5, pp. 173–176], etc.

What is of most interest to us in all this is Arzelà's theorem on the relationship between A_1 and B_1 convergence. To be specific, Arzelà proved that if a functional series $\sum_{n=1}^{\infty} u_n(x)$ converges in the sense of B_1, then its terms can be combined into groups

$$v_1(x) = \sum_{i=1}^{k_1} u_i(x), \quad v_2(x) = \sum_{j=k_1+1}^{k_2} u_j(x), \ldots,$$

$$v_p(x) = \sum_{p=k_{\nu-1}+1}^{k_\nu} u_p(x), \ldots,$$

(without changing the order of its terms) in such a way that the series $\sum_{m=1}^{\infty} v_m(x)$ converges in the sense of A_1. In other words, every functional series that converges uniformly on an interval in the sense of Dini can be transformed into a series that converges uniformly in the usual sense on the same interval by a suitable grouping of its terms, without changing their order.

A few years later this result was obtained by Hobson [1] independently of Arzelà. In his paper Hobson remarked that if each term of a series that converges in the A_1 sense is replaced by a sum of functions, then the resulting series will in general be divergent; if, however, it converges everywhere, then it must necessarily converge in the sense of B_1 [1, pp. 373–374]. From this he deduced that "the difference between uniform and simple uniform convergence is insignificant" [1, p. 374] and repeated this assertion much later [3, p. 107].

A propos of this inference, it is evidently useful to remark that because of terminological confusion in the question of uniform convergence the phrasing of Hobson's conclusion is unfortunate, since it does not stress the fact that the discussion refers only to the convergences A_1 and B_1. As we have seen, one of the B convergences makes it possible to establish a necessary and sufficient condition for continuity of the sum of a series at a point, while none of the A convergences allows this. Moreover, the relationships among the B convergences are more complicated than those among the A convergences. It may be, however, that Hobson's conclusion is not quite right per se.

3.4 Arzelà quasiuniform convergence

The Italian mathematician Arzelà is best known in mathematical literature in connection with a theorem on the compactness of families of functions which is actually due to his compatriot and contemporary Ascoli, as Arzelà himself pointed out more than once. Arzelà laid claim only to slightly different proofs and various applications of this theorem. However, his name is also remembered, and quite justifiably so, in connection with the concept of quasiuniform convergence, though attempts have been made to minimize his role in the study of this concept.

It was shown in the preceding sections that the uniform convergences of type A guarantee that the sum of an everywhere convergent series of continuous functions will be continuous, but that these convergences are not necessary for the continuity of the sum. The weakest of the convergences B, namely B_3, turns out to be both necessary and sufficient for the sum of such a series to be continuous at a point. But since convergence B_3 at each point of a closed set does not imply convergence B_1 in relation to the entire set, it turned out that none of the six kinds of convergence considered above made it possible to establish necessary and sufficient conditions for continuity of the sum on a closed interval or set as a whole. It was natural therefore that the search for such conditions continued.

Arzelà established these conditions in 1884, introducing yet another form of uniform convergence. We have not had access to the paper in which he did this,[22] but this is evidently the first appearance of the concept now known as Arzelà quasiuniform convergence. We shall study it on the basis of the much later long article of the same author [5].

---

[22] "Intorno alla continuità delle somme di infiniti funzioni continue," *Rend. dell'Accad. R. delle Sci. dell'Ist. di Bologna*, **19**, 1883–1884, pp. 79–84.

Here quasiuniform convergence, or, as Arzelà calls it, uniform convergence on closed intervals (convergenza uniforme a tratti), is not given a separate definition, as he had studied it earlier and considered it known. The definition, however, is actually contained in a theorem of Arzelà which we state in his own words:

> We assume that an infinite series of continuous functions has a definite sum $S(x)$ at every point of some interval $a\ldots b$. A necessary and sufficient condition for the sum to be finite and continuous is that for every arbitrarily given σ and every integer m_1 there exist another whole number $m_2 \geq m_1$ such that for some number m between m_1 and m_2 the inequality
>
> $$|R_m(x)| \leq \sigma$$
>
> holds, where $R_m(x)$ denotes the remainder of the series starting from the term $u_m(x)$; the number m may vary from one x to another [5, p. 153].

At this point Arzelà remarks that this necessary and sufficient condition amounts to a type of convergence different from simple uniform convergence and from Dini convergence and illustrated this by numerous examples [5, pp. 153–171].

If the definition of quasiuniform convergence contained in this statement is compared with the one usually given today,[23] it can be asserted that they coincide completely. Luzin was therefore mistaken when he wrote that "in attempting to make Arzelà's reasoning as simple and clear as possible, Borel subjected Arzelà's definition to a reasonable modification and thereby arrived at a new type of convergence that he named quasiuniform convergence [9, p. 256]. Luzin even gave Borel credit for the definition itself [9, p. 257] and called the theorem of Arzelà just given the Arzelà-Borel Theorem [9, p. 259]. In fact only the term *quasiuniform convergence* is due to Borel along with a few secondary clarifications. Borel himself [4, p. 44] did not claim anything else and definitely connected this theorem with Arzelà's name, referring specifically to Arzelà's paper [5].

The theorems of Dini and Arzelà gave a complete solution of the old problem of the continuity of the sum of a series of continuous functions, whether the convergence was at a point or on a closed set or interval. But many questions of the theory of functions were connected with the uniform convergences, and one of the most important of them was the question of integration of series. This operation is a very profound and multifaceted

---

[23] Cf., for example, Luzin [9, p. 257].

problem of the theory of functions, and even the study of one of the simplest integrals—that of Riemann—encountered difficult questions.

It is clear a priori that a given everywhere-convergent series, even a series of continuous functions, cannot necessarily be integrated termwise, not only in the sense of Cauchy, as Cauchy had asserted in 1823, but also in the sense of Riemann. Indeed as early as 1875 and 1881 respectively Smith [1] and Volterra [1] had constructed examples of pointwise-discontinuous functions that were not Riemann-integrable. But by Baire's theorem every such function is the sum of a convergent series of continuous functions. Consequently the limit of a sequence of functions, even continuous functions, is not in general Riemann integrable. This holds a fortiori for sequences of integrable functions, and one cannot therefore speak of the sum of the integrals as equal to the integral of the sum. The example of Darboux given in Sec. 2 above confirmed this in regard to a series of continuous functions converging to a series of continuous functions.

The operation of integration of series is one of the most common operations of analysis and the theory of functions, and for that reason it was studied and continues to be studied by many analysts. The number of papers on this subject is unimaginably huge, and we limit ourselves here to just a few connected with our thesis.[24]

As we have seen, in 1875 Darboux and Thomae established that convergence A_1 of a series of Riemann-integrable functions is sufficient for termwise integrability. In 1878 Dini found that convergence B_1 is sufficient for the sum to be integrable, but obtained no new conclusions from this convergence relating the integral of the sum to the sum of the integrals. In 1897 Bendixson obtained some very restrictive conditions for termwise integrability connected with convergence B_1.

After introducing quasiuniform convergence Arzelà also attempted to connect it with the question of termwise integrability. He began studying this question in 1885 and returned to it many times. He made it his goal to establish necessary and sufficient conditions for termwise integrability of a convergent series of Riemann-integrable functions; and although he did not succeed in general, his efforts in that direction yielded remarkable results.

He first took up this question in [2], published a year after the 1884 article mentioned above, in which the concept of quasiuniform convergence was first introduced. It may be that he first tried to use this kind of convergence, but almost immediately discovered that it was insufficient for his purposes,

---

[24] Many studies of integration from the past century and the beginning of the present one are studied in the book of Hawkins [1]; Rosenthal devoted a large section of his survey to this topic [1, pp. 1078–1085].

and in the note [1] he had to begin by proving an auxiliary proposition that can be stated as follows in modern language: If a series of functions converges everywhere, it also converges in measure.[25] He then introduces what is actually a new kind of uniform convergence, which he calls "general uniform convergence over intervals" (convergenza uniforme a tratti in generale), and which we shall call *general quasiuniform convergence*.

Arzelà's definition of the latter is as follows:

> A series $\sum_{1}^{\infty} u_n(x)$ that converges at every point between a and b is said to converge *in general over closed intervals* if for arbitrarily given positive numbers σ and ε and a finite number of segments $\tau_1, \tau_2, \ldots, \tau_p$ cut off between a and b of total length less than ε, for each integer m_1 there exists an integer $m_2 \geq m_1$, such that for all values of x between a and b except possibly those in the segments $\tau_1, \tau_2, \ldots, \tau_p$ there is an integer m between m_1 and m_2, possibly depending on x, such that the inequality
> $$|R_m(x)| < \sigma$$
> holds, where $R_m(x)$ is the remainder of the series [5, p. 326].

In other words general quasiuniform convergence is simply quasiuniform convergence neglecting a set of points that can be enclosed in a finite set of intervals of arbitrarily small length, i.e., neglecting a set that is discrete in the sense of Harnack. This kind of convergence, so far as I know, never received wide circulation in mathematical investigations, and it seems that only Hobson attempted to modify it and apply it to problems of integrating series [3, pp. 312–314]. However, Arzelà used it many times.

In the note just mentioned [2] he proved that general quasiuniform convergence of a series of Riemann-integrable functions to a bounded function is a necessary and sufficient condition for the sum to be Riemann-integrable. His proof contained a gap that will be discussed below. At present we note merely that in the reasoning he gave [2, pp. 323–324] he used Pincherle's theorem (mentioned on p. 97 above) without referring to Pincherle.

In another note of the same year [3], relying in particular on the theorem above on convergence in measure, he attempted to establish more: to find necessary and sufficient conditions for termwise integrability of a series of integrable functions. In analyzing the known examples of series for which the integral of the sum is not equal to the sum of the integrals, he noticed that they all had in common the fact that the sum of the indefinite integrals of the terms

---

[25] We shall return to this theorem in Sec. 6.

of the series is not a continuous function of the upper limit of integration, i.e., does not converge quasiuniformly. This led him to the erroneous conclusion that if the integral of the original series over the entire interval exists, then a necessary and sufficient condition for termwise integrability of the series is that the series of indefinite integrals of the terms converge quasiuniformly, i.e., to a continuous function. He attempted to prove this. His arguments are very complicated, and even Dini, who communicated this note for publication, did not notice that they were erroneous (assuming, of course, that Dini read it carefully).

Arzelà's paper [3] contains, besides this erroneous proposition, one essential result, namely a sufficient criterion for termwise integrability connected with the uniform boundedness of the remainder. Lebesgue [20, p. 125, footnote] ascribed this theorem to Osgood in the special case where the functions of the series and its sum are continuous. In fact Osgood did prove it in this form [1, pp. 176–182] in 1897. But Arzelà had proved it much earlier and for Riemann-integrable functions [3, p. 537], confirmed it [4] in 1897, and gave it a second proof [5, p. 723–725] in 1900.

The appearance of Osgood's paper [1] caused Arzelà to re-examine the fundamental result, as he regarded it, of his paper [3], since Osgood had constructed an example in which the indefinite integrals converge quasiuniformly,[26] but the integral of the sum is nevertheless not equal to the sum of the integrals [1, pp. 174–175]. His first reaction to Osgood's memoir was a note [4] in which he acknowledged his error. In this note he attempted to use general quasiuniform convergence fully but did not go beyond a theorem on termwise integration of series with uniformly bounded remainders and only promised to remove this restriction.

Arzelà kept his promise in the large and important memoir "Sulle serie di funzioni" [5] that received widespread recognition among mathematicians and to which we have already referred several times. This memoir contains a summary and re-examination of his fundamental results regarding the various kinds of uniform convergence, both those he had already published and those he had recently obtained.

In particular [5, pp. 706–712] he again gave a new proof of the theorem on necessary and sufficient conditions for the integrability of the sum of a series of integrable functions that converges to a bounded function. However, as in [3], his new proof contained a gap where he assumed that the set of points at which the partial sums of the series have a finite jump can be enclosed in a finite number of intervals with arbitrarily small total length [5, pp. 708–

---

[26] Osgood evidently did not know of Arzelà's work, and both his method and his terminology are different.

709]. Hobson removed this gap in 1904 and proved the theorem in a slightly different way [1, pp. 382–387]. Hobson [3, pp. 312–314] gave yet another proof of this proposition in 1926.

As for the sufficient criterion for termwise integrability of a series, Arzelà generalized it to the case when the requirement $|R_n(x)| < \varepsilon$ for all x and all n from some N on is weakened by allowing an at most countable set of exceptional values of x where this inequality does not hold [5, pp. 729–733]. For the case where all the functions are continuous Osgood gave the analogous generalization [1, pp. 187–188].

Arzelà also used quasiuniform convergence to establish sufficient conditions for termwise differentiability [5, pp. 725–727] and to study certain other questions. His attempt [5, pp. 177–182] after analyzing Ascoli's conditions for compactness to find a generalization of them is extremely interesting and perhaps still insufficiently appreciated. We recall that it was not until 1940 that Sirvint [1] proposed a generalization of the concept of compactness of families of functions based on Arzelà quasiuniform convergence.

It seems reasonable at this point to say a few words about a curious paper of the Italian Dell' Angola [1] published in 1907. Dell' Angola started from certain reasoning of Arzelà and probably did not know of the papers of Weierstrass [1] and Pincherle [1] that we discussed above in Sec. 2, in which a general form of the Heine-Borel theorem was studied along with some of its applications to questions of uniform convergence. Arzelà, as we mentioned, had also applied similar reasoning [3] as early as 1885. He had also used this reasoning, again not referring to Weierstrass or Pincherle, in [5, pp. 145–150] to establish necessary and sufficient conditions for continuity of the sum of a series of continuous functions on a closed interval. It was these arguments of Arzelà that led Dell' Angola to the two theorems of Pincherle discussed on p. 97 above. He again stated and proved them in almost the same form as Pincherle, then applied them to obtain a series of theorems of the theory of functions, the majority of which were unknown when Pincherle wrote his paper [1]. In particular Dell' Angola showed that the Heine-Borel theorem can be simply obtained from Pincherle's theorem for both countable and uncountable coverings [1, pp. 375–378]. He also gave a new proof of Arzelà's theorem on necessary and sufficient conditions for the continuity of the sum of a series and extended it to the complex case.

The concept of quasiuniform convergence introduced by Arzelà became an important tool of investigation in various mathematical questions. We limit ourselves to mentioning that in 1904 Hobson made use of it [1], having actually arrived at this concept independently of Arzelà; in 1905 Borel made brilliant use of it [4]; in 1905–1906 Fréchet introduced it into functional analysis by giving it a suitable generalization; in 1907 Montel [1] applied it in the theory of functions of a complex variable and Hobson [2] included it in his famous treatise on the theory of functions of a real variable; as mentioned, in 1940 Sirvint [1] discovered a new sphere of action for it in functional analysis; in 1948 it was given a wide generalization by Aleksandrov [3], etc. This list could easily be significantly enlarged, but there is no real need to do so.

3.5 Convergence almost everywhere

Convergence almost everywhere plays approximately the same role in the theory of functions that convergence everywhere plays in classical analysis. It differs from the latter, however, in that it has no single meaning, but depends on the concept of measure. Like convergence everywhere, convergence almost everywhere is frequently too broad for studying particular questions, and it is often restricted, while in solving other problems it is insufficient and must be generalized.

On the historical level this form of convergence, as far as I know, has not been studied at all and this first attempt of ours to do so will naturally err with all the defects characteristic of such attempts, especially in regard to completeness.

If it is possible to trace the sources of the uniform convergences with comparative completeness, and quasiuniform convergence can be confidently connected with the name of Arzelà, the situation is somewhat more complicated with almost everywhere convergence.

A sequence $\{f_n(x)\}$ is said to converge *almost everywhere* to a limiting function $f(x)$ if $f_n(x) \to f(x)$ in the ordinary sense except on a set of measure zero; at points of the latter set no restrictions are imposed on the behavior of $f_n(x)$; the sequence may diverge to infinity, or oscillate between several limits, it may be that nothing at all is known about its convergence or divergence, and so forth.

Rosenthal [1, p. 1079] ascribed this kind of convergence to Lebesgue and dated it to a 1910 paper [16], referring to a place where the phrase *convergence almost everywhere* is used [16, p. 376]. Although we basically agree with Rosenthal in crediting this concept to Lebesgue, we nevertheless place the date of its introduction a few years earlier, since Lebesgue was actually using this concept as early as 1901. It should be emphasized, however, that this dating is to some degree arbitrary, since reading the works of Lebesgue gives the impression that he did not regard this convergence as a new kind of convergence. In his writings it is sometimes difficult to distinguish convergence almost everywhere from convergence everywhere, since he usually did not specify the hypothesis that sets of measure zero could be neglected.

3.5 Convergence almost everywhere

The question of the evolution of this concept is closely connected with the more general question of the evolution of the principle of negligibility of sets of points of various types in many different problems of the theory of functions. Not to delve to deeply into the past, one can date the conscious application of the latter principle to 1870, when, as we have seen on p. 92 above, Heine began to do this in relation to finite sets. In the following year Cantor took the same route, proving that the uniqueness theorem for trigonometric series remains valid if we neglect any finite set of points. A year later Cantor carried out a radical extension of this principle, generalizing the uniqueness theorem to infinite sets P such that $P^{(\nu)} = 0$. Throughout the book of Dini [4] the principle of neglecting sets of the latter form was applied in many questions, in particular in questions of convergence of series. We shall give one of Dini's theorems as an example.

If the series $\sum u_n$ converges at all points of a finite interval (α, β) except possibly a set G of points that is finite or infinite of the first kind, and at whose points all or some of the terms become infinite or undefined; if at the same time these terms $u_1, u_2, \ldots, u_n, \ldots$, are integrable between α and β when some values are assigned at these points of indeterminacy, and the series of integrals $\sum_{1}^{\infty} \int_{\alpha}^{\beta} u_n\, dx$ converges and is a finite and continuous function between α and β, and if the series $\sum_{1}^{\infty} u_n$ is termwise integrable, then the sum of the series $\sum_{1}^{\infty} u_n$ is integrable over the interval (α, β) when any values are assigned to it at points of indeterminacy and the relation

$$\int_{\alpha}^{x} dx \sum_{1}^{\infty} u_n = \sum_{1}^{\infty} \int_{\alpha}^{x} u_n\, dx$$

holds for each value of x between α and β (including α and β) [4, p. 390].

Harnack proceeded in a similar way starting in 1881 in relation to the so-called discrete sets,[27] as did Arzelà after him in the results we have already discussed, and as many other authors did. After introducing the notion of a set of first category, Baire [3; 5, p. 65] began to ignore these sets also in many questions.

Thus by the beginning of the twentieth century the principle of neglecting various classes of sets had become quite widespread.

In 1901 Lebesgue [2, p. 1027] states this principle in passing in relation to sets of measure zero in stating a theorem on necessary and sufficient conditions

---

[27] Cf. Medvedev [2, pp. 113–115], Hawkins [1, pp. 58–61].

for Riemann integrability of a function. In this same paper he states the following theorem: *If a sequence of integrable functions has a limit, then that limit is integrable.*[28]

No one nowadays doubts that in this formulation it is really convergence almost everywhere that is being discussed, the more so since the principle of negligibility of sets of measure zero, as we have said, is found on the same page, though in application to another theorem. Even so, reading this passage gives grounds for doubt: it relies on results of Baire based on convergence everywhere.

Such an ambiguous position is preserved in Lebesgue's 1902 dissertation. The verbal statement of the theorem remains almost the same, and it is clear from the proof [3, p. 257] that convergence almost everywhere is at least not excluded; but in the description of the class of measurable functions [3, pp. 257–258] the results of Baire are again invoked. There is also no doubt nowadays that in the statement of sufficient conditions for integrability of a sequence of functions [3, p. 269] it is convergence almost everywhere that is meant. Nevertheless the footnote on the same page can be disturbing: Lebesgue notes that Osgood had proved a special case of the theorem in 1897, and he says that the latter had proved the theorem for continuous functions (as terms of the series and as limit), but he does not point out that Osgood had in mind convergence everywhere,[29] and that he was using a different kind of convergence.

Probably the most interesting paper in this respect is Lebesgue's 1903 note [5]. A reading of Baire's dissertation had given Lebesgue the idea that the principle of negligibility of various classes of sets was both possible and reasonable, not only sets of measure zero but also sets of arbitrarily small measure, depending on the particular problem under consideration. It is here [5, pp. 1228–1229] that this principle is stated in its general form. In the aspect we are now interested in, the most important of Lebesgue's statements is the following: *Every convergent series of measurable functions is uniformly convergent when certain sets of measure ε are neglected, where ε can be as small as desired* [5, p. 1229].

Certain considerations immediately arise from comparing this statement with the statement of Egorov's Theorem given in standard textbooks: *Suppose that on a measurable set E we are given a sequence of measurable functions $f_1(x)$, $f_2(x)$, $f_3(x),\ldots$ that are finite almost everywhere and converge almost*

---

[28] At the time Lebesgue did not distinguish between integrable and measurable functions. This theorem actually refers to bounded measurable functions.

[29] However, as we have said on p. 109 above, Osgood had also proved it neglecting a countable set.

everywhere to a finite function

$$\lim_{n \to \infty} f_n(x) = f(x). \quad (*)$$

Then for any $\delta > 0$ there exists a measurable set $E_\delta \subset E$ such that

1) $mE_\delta > mE - \delta$,

2) *on the set E_δ the limit $(*)$ is approached uniformly* (Natanson [1, Vol. I, pp. 99–100]).

It is clear first of all that Lebesgue's assertion contains Egorov's Theorem. Ignoring inessential details, we can say that the main difference between these two propositions lies in the understanding of the kind of convergence. If Lebesgue meant only convergence everywhere in the hypothesis, then the theorem he states is a special case of Egorov's Theorem; but if he understood it as convergence almost everywhere, then the two statements coincide. Lebesgue himself again did not explicitly indicate in [5] what he meant by "any convergent series of functions." But taking into account the fact that it was here that he enunciated in full generality the principle of the negligibility of various classes of sets, and that not only in [5] but in his other papers up to 1910 he does not seem to have emphasized the distinction between convergence everywhere and convergence almost everywhere, we may suppose that it was convergence almost everywhere that he had in mind. This is the way Severini [1] perceived it in 1910 when he proved the special case of Egorov's theorem that occurs when the sequence $\{f_n(x)\}$ is not an arbitrary sequence of measurable almost-everywhere finite functions but a sequence of orthogonal functions.[30] Egorov assumes convergence almost everywhere from the very beginning, although his way of expressing himself, characteristic of mathematicians of the time—"we assume a sequence of measurable functions...converging to a limiting function $f(x)$ for all points x of the interval AB except possibly the points of a set of measure zero" [1, p. 244]—testifies that there was still no clear boundary between the two forms of convergence.[31]

Thus it can be asserted that Lebesgue not only was using the concept of almost everywhere convergence in 1903, but also stated Egorov's Theorem, which characterizes this kind of convergence.

---

[30] It should be noted, however, that Severini does not use the fact that the $f_n(x)$ are orthogonal functions in his proof; his reasoning applies in the general case.

[31] Similarly Severini speaks of convergence "at all points except at most the points of a set of measure zero" [1, p. 3].

114   Chapter 3   Sequences of functions. Various kinds of convergence

Nevertheless in his subsequent papers he continues to conflate convergence everywhere with convergence almost everywhere, and the two are distinguished only by context. In his book on integration and primitives [8], published in 1904, the same theorem on the limit of a sequence of measurable functions is stated and proved in its previous form with the same explanations about the extent of the class of measurable functions [8, p. 111]; to be sure, these explanations are broadened by the application of the principle of negligibility of sets of measure zero in the proof that Riemann integrable functions belong to this class [8, p. 112], but this principle is still not connected with convergence. Moreover, as we shall see in the next section, in his book [12] Lebesgue states one theorem only for convergence everywhere, although it is fully applicable in the case of convergence almost everywhere (see p. 121).

It is difficult to say which mathematician, in considering questions of convergence, was the first to add the words "except possibly for a set of measure zero" or the equivalent to the assertion that a series or sequence converges. In any case Fréchet definitely did this in 1905 [2, p. 38]; he does the same in 1906 [4, pp. 15–17]; the situation is similar with F. Riesz in 1907 [2, 4], etc. The term *almost everywhere* can be found in a 1909 paper of Lebesgue [14, p. 43].

Of course all these details might seem unimportant to a modern reader, since we can now trace very clearly the idea of convergence almost everywhere in papers in which the authors did not distinguish it from the idea of convergence everywhere.[32] But while Fréchet or Riesz spoke of the former as of something already known in the first decade of the century, the situation is different with Weyl, and it may be assumed that in 1909 he was unacquainted with it. Indeed in his paper on orthogonal series [1] he introduced the concept of *essentially uniform* (wesentlich gleichmäßig) convergence of a series of functions in the following words:

I say that the series

$$u_1(x) + u_2(x) + \cdots, \qquad (1)$$

where $u_1(x), u_2(x), \ldots$ are any functions defined on $0 \leq x \leq 1$, converges *essentially uniformly* if for every positive number $\varepsilon < 1$ there exists a set U_ε of measure $1 - \varepsilon$ contained in the interval

---

[32] However, even today this is conjectural in some cases. Thus, for example, Bari and Men'shov [1, pp. 389–390], commenting on Luzin's words in his dissertation that "one can establish that there exist convergent trigonometric series whose sum is not integrable in the sense of Denjoy on any interval, no matter how small" (Luzin [4, p. 54]), surmise only as a conjecture that Luzin had in mind convergence almost everywhere.

0 ... 1 such that for all values of x belonging to U_ε the series (1) converges uniformly [1, p. 225].

With this definition Weyl introduced into the theory of functions a kind of convergence more general than convergence almost everywhere and now known as almost uniform convergence (Halmos [1, p. 89]).[33] But he could hardly have suspected this at the time. His basic results referred to measurable functions defined on a set of finite measure, and for sequences of such functions convergence almost everywhere and essentially uniform convergence coincide. One may conjecture that if Weyl had known about convergence almost everywhere, he would probably not have introduced essentially uniform convergence, since the latter was not needed in his research. All the main results of Weyl's paper have been perceived as relating precisely to convergence almost everywhere.

Of course convergence almost everywhere is a generalization of convergence everywhere that seems almost natural when the principle of the negligibility of sets of measure zero is understood. Nevertheless these two types of convergence are profoundly different. It suffices to recall that every sequence of continuous functions that converges everywhere yields a function of Baire class 1 at worst, and even the transfinite iteration of such limiting passages does not yield results outside the class of Borel measurable functions. In contrast, even a single limiting passage almost everywhere makes it possible to obtain any measurable almost everywhere finite function starting from a sequence of rather simple functions. Thus in 1906 Fréchet proved that for every measurable and almost everywhere finite function defined on a closed interval there exists a sequence of continuous functions converging to it almost everywhere [4, pp. 15–16].[34] In 1920 Riesz [11, p. 209] essentially established that every measurable and almost everywhere finite function can be represented as the limit of a sequence of step functions. Men'shov [2] arrived at this same result in 1941 using the sequence of partial sums of a trigonometric series instead of a sequence of step functions. The latter result is interesting in the present context because not every continuous function can be represented by a trigonometric series that converges to it at every point.

Consequently convergence almost everywhere is immensely more powerful than convergence everywhere as a means of representing a function an-

---

[33] As Halmos correctly remarked [1, p. 89], the term *almost uniform convergence* is an unfortunate one. Weyl's term was more suitable; but most likely because the convergence he introduced was perceived as convergence almost everywhere, his terminology has not survived.

[34] An earlier statement of this theorem by Fréchet [2, p. 28] referred to functions belonging to the Baire classification. We shall not go into detail about the more general sense of Fréchet's reasoning that led him to this formulation.

alytically. And where repeated application of convergence everywhere did not lead beyond the class of pointwise-discontinuous functions—the principal object of study in the theory of functions of a real variable during the nineteenth century—convergence almost everywhere made it possible to study the totality of all measurable almost everywhere finite functions.

Naturally in studying the more specific questions of the theory of functions such a general kind of convergence was of necessity restricted. Some particular restrictions were motivated by the inadequate level of development of the subject; such was the situation, for example, in the case of Dini and Harnack, when the concept of measure zero in the sense of Borel and Lebesgue did not yet exist,[35] and they were simply forced to neglect only less general sets of measure zero in studying questions of convergence. In contrast, in the question of the uniqueness of the representation of functions by trigonometric series, for example, such restrictions were inherent in the very essence of the matter. Men'shov's famous example [1] of a trigonometric series with nonzero coefficients that converges almost everywhere to zero showed that in this problem one cannot ignore a general set of measure zero, so that convergence almost everywhere is demonstrably too broad; to what extent it needs to be strengthened is unknown even today.

As with convergence everywhere, various restrictions have been imposed on convergence almost everywhere in the sense of uniformity. The best known of these convergences is naturally convergence of type A_1, which received the name of uniform convergence almost everywhere (Halmos [1, p. 88]). However, as far as I know, for uniform convergence almost everywhere no work has been carried out similar to that described in the three preceding sections in relation to the various kinds of uniform convergence everywhere.

In speaking of convergence almost everywhere up to now we have meant the negligibility of sets of measure zero in the sense in which they were understood by Borel and Lebesgue themselves, i.e., sets that can be enclosed in a finite or countable set of intervals of arbitrarily small size (length, area, volume, etc.) in a Euclidean space. One can, however, use more general definitions of measure and present the negligibility of sets of measure zero as a principle with a more general interpretation. Historically such an extension first occurred in relation to Lebesgue-Stieltjes measure and was carried out in the works of Radon [1], Young [5, 6], and Daniell [1, 2].

The definition of convergence almost everywhere with respect to a Lebesgue-Stieltjes measure is verbally the same as that given above on p. 110, but a set of measure zero is understood in the sense of the new definition of measure. But this external similarity conceals an important difference in

---

[35] For the introduction of this concept see, for example, Hawkins [1, pp. 97–106, 122–124].

the definition of measure and the consequent forms of convergence dependent on it. A given set of a Euclidean space always has only one Lebesgue measure (if it has any measure at all). But one can assign to that set as many different Lebesgue-Stieltjes measures as desired. Even with a fixed function defining the measure convergence almost everywhere in the sense of Lebesgue and almost everywhere in the sense of Lebesgue-Stieltjes may turn out to be independent of each other, since a set may have Lebesgue measure zero and positive Lebesgue-Stieltjes measure. Therefore for convenience in exposition we shall call the former *L-convergence almost everywhere* and the latter *L-S-convergence almost everywhere*.

The situation regarding the introduction of *L-S*-convergence almost everywhere is even more complicated than that regarding *L*-convergence almost everywhere. Lebesgue-Stieltjes measure itself was introduced in general form by Radon [1] in 1913. Naturally the general concept of a set of *L-S*-measure zero and the principle of negligibility of such a set in various questions, including convergence questions, first appears in his writings. However, while Lebesgue, though not emphasizing the distinction between convergence everywhere and *L*-convergence almost everywhere, nevertheless occasionally spoke of a principle of negligibility of sets of *L*-measure zero, Radon did not state any analogous principle in [1]. Moreover when he stated theorems on convergence, he either did not state this principle explicitly and simply referred to their analogues in the Lebesgue theory, or spoke of convergence without specifying what he meant. We shall give two examples.

After defining the concept of a function measurable with respect to *L-S*-measure, Radon continues:

> For functions measurable with respect to f, by the same methods as in the theory of functions measurable in the sense of Lebesgue, one can prove a variety of theorems expressing the fact that the class of functions measurable with respect to f contains as its elements functions obtained by elementary arithmetic operations, and the limiting processes do not lead outside of this class. The latter can be stated as follows:[36]
>
> If $F_1, F_2, \ldots$ are measurable with respect to f, then both of the limits of indeterminacy
>
> $$\overline{\lim_{n=\infty}} F_n, \underline{\lim_{n=\infty}} F_n,$$
>
> are measurable provided the points at which, for example, the relation $\overline{\lim}_{n=\infty} F_n = \infty$ holds are always among the points at which

---

[36] Cf. Vallée-Poussin, *Cours d'analyse infinitesimale*, T. II (Paris, 1912) [Radon's footnote].

the value of the function is larger than the number A, no matter how large the latter is taken [1, p. 1325].

Of course all this is true not only for Radon-measurable functions, but also in the study of more general definitions of measure. But to see in this, for example, the theorem that an almost-everywhere convergent sequence of functions that are L-S measurable in this sense has a limit that is measurable in the same sense is not easy, especially since the only proof given is what was quoted above. The reference to Vallée-Poussin only increases the difficulty, since the latter arrived at the concept of Lebesgue-Stieltjes measure [2] only in 1916 independently of Radon and had apparently not suspected it earlier; in 1912, in the place referred to by Radon, Vallée-Poussin was talking only about Lebesgue measure.

Here is another formulation of Radon's [1, p. 1330]:

If $F_1, F_2, \ldots$, is a convergent sequence of functions that are measurable with respect to f and $|F_n| < M$, then

$$\lim_{n=\infty} \int_E F_n \, df = \int_E (\lim_{n=\infty} F_n) \, df.$$

Although this statement mentions the convergence of the sequence, nothing can be gleaned either from the statement or from the remarks on its proof about the kind of convergence.

In other words Radon made even less distinction than Lebesgue between convergence everywhere and almost everywhere, at least in this paper.

W. H. Young, in introducing Lebesgue-Stieltjes measure in 1914 (also independently of Radon), considered it necessary to emphasize the specifics of a set of measure zero in the new definition of measure, exhibiting an example of a set of Lebesgue measure zero but of positive Lebesgue-Stieltjes measure [5, p. 133]; he emphasized this even more insistently in 1916 [6, p. 55, footnote]. The principle of negligibility of sets of L-S-measure zero was widely used in both of these papers in various questions of differentiation and integration, but not in the questions of convergence that we are interested in. Daniell [1,2] proceeds in the same way in 1918–1919.

It is evidently not a coincidence that Radon spoke rather vaguely of convergence L-S-almost everywhere and W. H. Young and Daniell attempted to avoid it. The fact is that Egorov's theorem plays a fundamental role in questions of convergence L-almost everywhere. But since a set of L-measure zero may have positive L-S-measure, the variant of Egorov's theorem discussed above is useless here, and this theorem had to be generalized. These were the considerations that motivated the work of Riesz [10, p. 221].

If we adapt his considerations to the language we are using, the gist of them is as follows. He introduces Lebesgue-Stieltjes measure [10, p. 221], defines the concept of a function measurable with respect to this measure [10, p. 222], and shows that if a sequence of functions measurable in this sense and almost everywhere finite converges L-S-almost everywhere on $[a, b]$, then it converges uniformly on a set whose complement relative to $[a, b]$ has arbitrarily small L-S-measure [10, pp. 222–223].

It would require too long and complicated a discussion to study the subsequent modifications in the concept of convergence almost everywhere and their applications in many different kinds of research, since this would lead us far beyond the bounds of the theory of functions proper as outlined above. Even in the case of Riesz' paper [10] we were forced to depart from the author's language, since otherwise we should have been obliged to invoke facts related more to the history of functional analysis than to the history of the theory of functions. We shall say a few more words more about two variants of Egorov's Theorem, but only as illustrative examples.

Egorov's Theorem was stated by Carathéodory [3] in 1940 for convergence almost everywhere with respect to a still more general measure. In his terminology it reads as follows:

> If a sequence of place-functions (Ortsfunktionen) $f_1, f_2, \ldots$ converges on a *soma* M for which $\varphi(M) < +\infty$, then to every positive number ε one can assign some part M_ε of the *soma* M such that, first, $\varphi(M_\varepsilon) < \varepsilon$ and, second, the sequence f_i converges uniformly on $(M - M_\varepsilon)$ [3, p. 184].

To keep from departing too far from our terminology and at the same time introducing long explanations connected with the language of Carathéodory, we shall make only the following remark. For Carathédory a *soma* is an analogue of a set (in general it is an element of a boolean algebra); the function $\varphi(M)$, which he called the *soma function*, is a generalization of Lebesgue-Stieltjes measure; a place-function is an analogue of a point function; convergence of $\{f_i\}$ to f means convergence almost everywhere with respect to the measure φ.

Finally, as a concluding example we mention the generalization of Korovkin [1], who, having noticed that not all the properties of ordinary measure were used in the proof of Egorov's Theorem, essentially introduced the negligibility of sets more general than sets of measure zero in the most general measure theory[37] and proved a Egorov Theorem for this type of convergence.

---

[37] He did not define this convergence explicitly; it is contained in his statement and proof of the generalized Egorov Theorem.

To illustrate the generality of his result Korovkin remarked that in particular a set function such as capacity, not having the property of additivity and therefore not being a measure in the traditional sense, can be taken as the general "measure" he had introduced, so that convergence almost everywhere with respect to a capacity can also be regarded as uniform neglecting a set of arbitrarily small capacity.

3.6 Convergence in measure

The evolution of the concept of convergence in measure is connected with the study of the question of termwise integration of series, more specifically with a sufficient condition for termwise integrability based on the uniform boundedness of the remainders of the series. This connection in the research of the nineteenth century, which goes back to Kronecker (1879), has been studied by Hawkins [1, pp. 111–119], and we shall go into detail only on the result of Arzelà mentioned above on p. 107 which Arzelà needed precisely for the study of conditions for termwise integrability of series of functions.

The point of departure for Arzelà's many studies was a lemma of geometrical content, which he stated as follows.

Suppose that a closed interval $[a,b]$ is given on the x-axis of a rectangular coordinate system and a sequence of points $\{y_s\}$ converging to the point y_0 is given on the y-axis; straight lines parallel to the x-axis are drawn through the points y_s, and pairwise disjoint intervals $\delta_{i,s}$ are marked off on these lines inside a rectangle bounded by the lines $x = a$, $x = b$, $y = 0$, $y = y_0$ in such a way that the number of such intervals on each of these lines inside the rectangle is finite, but possibly different from one line to another, and may increase without bound as y_s approaches y_0. Finally, suppose the sum of the lengths of the intervals $\delta_{i,s}$ on the line $y = y_s$ is d_s.

Then, if the condition $d_s \geq d$ always holds for every value of $s = 1, 2, \ldots$, where d is a fixed positive number, it follows that there exists at least one point $x = x_0$ on $[a,b]$ such that the line through x_0 perpendicular to the x-axis intersects infinitely many of the intervals $\delta_{i,s}$.

When the segments $\delta_{i,s}$ are projected orthogonally to the x-axis, which Arzelà seems not to have done explicitly, this lemma assumes the following form: *Given a countable sequence $D_1, D_2, \ldots, D_i, \ldots$ of systems of pairwise disjoint intervals $\delta_{i,s}$ contained in the interval $[a,b]$ such that the sum of the lengths of the intervals of each system d_s is at least d, there exists a point $x = x_0$ on $[a,b]$ belonging to an infinite number of the systems D_i.*

Arzelà [1] proved this lemma 1885 and proved it again in the paper [5, pp. 130–134] that we have mentioned many times. Subsequently many mathematicians turned their attention to this lemma in its many variant forms:

Borel [3], W. H. Young [2], G. M. Fikhtengol'ts [1], Riesz [9], Bieberbach [1], Landau [1], and others. Moreover while W. H. Young, G. M. Fikhtengol'ts, Riesz, and Landau connected it with the name of Arzelà, Borel and Bieberbach presented it as their own discovery. While Borel had a certain right to do so since he had stated it in general form, as W. H. Young was to do independently somewhat later, Bieberbach used it practically in Arzelà's formulation and applied it to the particular case of Arzelà's theorem on termwise integration of series.

As one of the corollaries of his lemma Arzelà stated the following proposition: *If $\sum_{1}^{\infty} u_n(x)$ is a series of functions that converges at every point of the interval $b - a$, then the sum of the lengths of certain intervals at each point of which the inequality $|R_n(x)| > \sigma$ holds for a given value of n must decrease without bound as n grows without bound* [1, p. 267].

He pointed this out again in his 1900 paper [5, p. 135].

It is not difficult to discern in this proposition an early variant of the modern theorem known as Lebesgue's theorem that if a series converges almost everywhere, then it converges in measure also. Arzelà stated it only for convergence everywhere and for outer Peano-Jordan measure.

Lebesgue evidently did not know these works of Arzelà. In his 1902 paper [3, p. 259] he used this theorem to find conditions for termwise integrability of a series[38] without stating it in explicit form. The next year he states it explicitly [5, p. 1229], but does not point out what kind of convergence he has in mind, convergence everywhere or almost everywhere, although the context suggests the latter. However, when Lebesgue again turns to this theorem in 1906 and gives a proof of it, the statement definitely indicates convergence everywhere: *If a series of measurable functions converges at all points of an interval, then the points of the interval for which one of the remainders from n on is larger than the quantity $\varepsilon > 0$ has arbitrarily small measure if n is taken sufficiently large* [12, p. 10].[39] Borel had stated and proved this theorem in just this way for convergence everywhere a year earlier [4, p. 37], and in his proof convergence at every point is essential; as for Lebesgue's proof of 1906, it is not entirely correct.[40]

It seems that, until Riesz [5, p. 396] made the extension in 1909, no

---

[38] Arzelà [2] had done this also as early as 1885, but for series of Riemann-integrable functions.

[39] Among other things the formulations of Arzelà and Lebesgue given here differ in that the latter requires that the inequality $|R_n(x)| \geq \varepsilon$ hold only for at least one remainder starting from the nth. Egorov [1, pp. 244–245] showed that this difference is unimportant.

[40] Cf. Riesz [5, p. 396].

one extended this theorem to convergence almost everywhere via the simple observation that, since it is a question of the measure of the set of points at which $|R_n(x)| \geq \varepsilon$, the truth of the assertion is unaffected by divergence on a set of measure zero. Since that time (and probably even earlier) despite the literal formulations of Borel and Lebesgue, this theorem has been taken by mathematicians to be precisely a theorem on convergence almost everywhere. Egorov [1, p. 244] proceeds this way in 1911, although he refers to the above-mentioned places in the works of Borel and Lebesgue. The theorem itself is frequently called Lebesgue's Theorem,[41] although it could with equal justice be connected with the names of Arzelà, Borel, and perhaps Riesz. Moreover, evidently not knowing about the works discussed above, Weyl arrives at essentially the same theorem in 1909, but in a somewhat less general and, so to speak, inverted form [1, pp. 229–230]. His approach is highly original, and we shall give some more details about it.

Weyl considers a sequence $\{f_n(x)\}$ of functions defined on $[0, 1]$. Denoting the largest of the numbers $|f_1(x)|, |f_2(x)|, \ldots, |f_n(x)|$ by $\tilde{f}_n(x)$ for each $x \in [0, 1]$, he concludes that $|f_m(x)| \leq \tilde{f}_n(x)$ for each $m \leq n$ and that $0 \leq \tilde{f}_1(x) \leq \tilde{f}_2(x) \leq \cdots$. Then taking an arbitrary nondecreasing function $\sigma(l)$ that is defined for $l \geq 0$ and positive for $l > 0$, he restricts the original sequence by the requirement that the function $\sigma(|f_n(x)|)$ be Lebesgue integrable, from which it follows that the integrals

$$\int_0^1 \sigma(\tilde{f}_n(x))\, dx = I_n$$

exist and that the sequence of these integrals converges to a finite limit I. As a fundamental lemma in his researches in [1] Weyl proves the following proposition: *If l is a positive number, then the values of x for which all the equalities*

$$|f_n(x)| \leq l \quad (n = 1, 2, \ldots),$$

hold form a set whose measure is at most $1 - \dfrac{I}{\sigma(l)}$ [1, p. 230].

It is not difficult to see the connection between Weyl's lemma and the theorem under consideration: instead of the measure of the set $E(|f_n - f| \geq \sigma)$ he estimates the measure of the set $E(|f_n| \leq \sigma)$, showing that the latter tends to the measure of the set E if the functions of the sequence satisfy the restrictions imposed on them. We recall that in this paper he also arrived at the concept of convergence almost everywhere independently (see pp. 114–115 above) and in an equally original way.

---

[41] Cf., for example, Natanson [1, Vol. I, p. 95].

3.6 Convergence in measure

We have given so much attention to this theorem partly because its history shows clearly those fluctuations in relation to the concepts of everywhere and almost everywhere convergence discussed in the preceding section, but mainly because it was this theorem that gave rise to the notion of convergence in measure. The latter is now defined as follows:

Suppose that on a set E there is defined a sequence of measurable, almost everywhere finite functions

$$f_1(x), f_2(x), \ldots, f_n(x), \ldots \tag{1}$$

and a measurable almost everywhere finite function $f(x)$. If it happens that

$$\lim_{n \to \infty} [mE(|f_n - f| \geq \sigma)] = 0$$

for any positive number σ, we say that the sequence (1) *converges in measure* to $f(x)$. Here the symbol $mE(|f_n - f| \geq \sigma)$ denotes the measure of the set of points $x \in E$ at which $|f_n(x) - f(x)| \geq \sigma$.

The theorem discussed above can be stated in the following form. Suppose on a measurable set E there is given a sequence of measurable and almost everywhere finite functions $f_1(x), f_2(x), \ldots, f_n(x)$ that converges almost everywhere on E to an almost everywhere finite function $f(x)$. Then for any $\sigma > 0$ the relation

$$\lim_{n \to \infty} [mE(|f_n - f| \geq \sigma)] = 0$$

holds.[42]

Consequently, taking the preceding definition into account, we can state the theorem more briefly as follows: if a sequence of measurable almost everywhere finite functions converges almost everywhere, then it also converges in measure. In other words, the kind of convergence just defined is essentially contained in this theorem and has been applied in mathematics at least since 1885 and to a certain degree even since 1879, although with certain restrictions, to be sure. But it was not until 1909 that Riesz [5] introduced it as an independent kind of convergence of a sequence of functions and interpreted the preceding theorem in this sense.

In the same memoir Riesz established a criterion for convergence in measure of a sequence of functions: *Let $\{f_n(x)\}$ be a sequence of measurable functions; in order for a function $f(x)$ to exist that is the limit in measure of the sequence $\{f_n(x)\}$ it is necessary and sufficient that the relation*

$$\lim_{\substack{p \to \infty \\ q \to \infty}} [mE(|f_p - f_q| \geq \sigma)] = 0 \tag{2}$$

---

[42] Cf., for example, Natanson [1, Vol. I, pp. 95–96].

hold for any $\sigma > 0$ [5, p. 396]. This criterion is applicable only when the set E has finite measure. For a set of finite measure Eq. (2) guarantees the existence of a unique limit $f(x)$ (up to a function equivalent to zero) to which the sequence converges in measure. We shall not go into detail about the question of convergence in measure on sets of infinite measure, referring the reader to the book of Hobson [3, pp. 239–245] and giving only the warning that convergence in measure is called convergence in mean in Hobson's book, and the references to the papers of Fischer and Riesz are wrong (Hobson [3, p. 240]). The incorrect terminology and references were caused by the fact that Hobson conflated two types of convergence that are usually distinguished in the mathematical literature—convergence in measure and convergence in mean.[43] The first of these is more general than the second, and this apparently tempted Hobson to conflate the two, causing also the incorrect reference to the works of Fischer and Riesz whose subject was convergence in mean. It is hardly rational to conflate these two kinds of convergence, since this leads to even greater difficulties than confusing convergence everywhere with convergence almost everywhere.

Riesz demonstrated the importance of the new kind of convergence with many examples. Erroneously assuming that the Fourier series of a square-integrable function is in general divergent almost everywhere[44] he stated the correct proposition that the Fourier series of every such function converges in measure (Riesz [5, p. 396]). He went on to show that Fatou's Lemma on passage to the limit under the integral sign can be generalized from convergence almost everywhere to convergence in measure.

The most important of Riesz' theorems, however, was the theorem that every sequence $\{f_n(x)\}$ that converges in measure to a function $f(x)$ contains a subsequence $\{f_{n_k}(x)\}$ that converges to $f(x)$ almost everywhere; Riesz proved this theorem in the same paper. It is in a certain sense an incomplete converse of the theorem discussed at the beginning of this section, and in that respect it recalls Egorov's theorem by the way in which it links convergence in measure with convergence almost everywhere.

Of course, just as convergence almost everywhere can be generalized in applying more general definitions of measure, so also convergence in measure can be broadened if we use concepts of measure different from Lebesgue measure proper.[45] We shall not dwell on this, but illustrate various applications

---

[43] We remark that there is no single term for the latter kind of convergence; for this see the next section.

[44] This assumption is true for integrable functions, as Kolmogorov [1] showed in 1926; the Fourier series of an integrable function may diverge at every point.

[45] Cf., for example, Halmos [1, pp. 90–94].

of this kind of convergence.

In 1912 Borel stated an important theorem: *Suppose given a bounded function F defined analytically[46] and two numbers ε and α; then there exists a polynomial P such that the set of points at which the absolute value of the difference between F and P exceeds ε can be enclosed in elementary sets[47] of total measure at most α.*[48]

Relying on this theorem and not referring to Riesz, Borel introduced the concept of convergence in measure, calling it *asymptotic convergence*. He then applied this convergence to obtain a conceptually very simple definition of the concept of the integral for bounded functions. The integral of a polynomial is assumed known; the integral of $f(x)$ is defined as the limit of the integrals of a sequence of polynomials converging to $f(x)$ in measure. Borel gave a detailed exposition of this integral in 1912 [7] and 1914 [8].

As we have said more than once, one of the important purposes being pursued in introducing the various forms of convergence was the representation of functions of the largest possible class using limits of sequences of functions of a given class (or sums of series of such functions). In this respect convergence in measure is in a certain sense unsurpassed.

Just from the theorem of Borel presented here it follows almost immediately that every measurable almost everywhere finite function is representable as the limit in measure of some sequence of continuous functions.[49] To be sure, for measurable almost everywhere finite functions convergence almost everywhere can play a similar role, as discussed in the preceding section. But while the class of measurable almost everywhere finite functions basically satisfied the needs of mathematicians in the first quarter of the twentieth century, the interest of mathematicians afterward came to be drawn more and more often to functions assuming infinite values on sets of positive measure. A weighty consideration in all this was on the one hand the fact that the definition of a measurable function by no means restricts the function to be finite almost everywhere. On the other hand, in the first half of the twentieth century many examples of functions assuming infinite values on sets of positive measure were constructed. This posed the problem for analysts of studying such functions, in particular, finding a way of representing them analytically. This last problem turned out to be not so simple, and although Luzin [4, p. 190] had posed it as early as 1915, more than three decades were to pass before Men'shov proposed the first solution to it. Before discussing this proposition,

---

[46] A function belonging to the Baire classification is meant.
[47] Here Borel defines an elementary set to be an interval.
[48] For a modern formulation see Natanson [1, Vol. I, pp. 110–111].
[49] Cf. Natanson [1, p. 104].

we should return to the definition of the concept of convergence in measure stated earlier.

In the definition of this kind of convergence given on p. 123 it was explicitly stated that the subject was a sequence of measurable almost everywhere finite functions converging to a function that was also measurable and finite almost everywhere. But if we return to earlier formulations, we see that the situation is not quite like this. Riesz himself, for example, gives the definition as follows [5, p. 396]:

> Let $f_1(x), f_2(x), \ldots$ and $f(x)$ be measurable functions defined on a set E; let ε be some positive number; we denote by $m(n, \varepsilon)$ the measure of the set
> $$[|f(x) - f_n(x)| > \varepsilon].$$
> Then we shall say that the sequence $[f_n(x)]$ *converges in measure* to the function $f(x)$ if the relation
> $$\lim_{n=\infty} m(n, \varepsilon) = 0$$
> holds for any quantity ε.

Although only measurable functions are mentioned here, it is clear from the context that Riesz has in mind only measurable almost everywhere finite functions.

In his short note [6] of 1912 Borel introduced convergence in measure only for sequences of polynomials. In 1914 he first studied it for bounded functions [8, pp. 242–243], and then extended it to almost everywhere finite functions [8, p. 248], not because he had consciously set the goal of restricting the definition of the concept of convergence in measure, but simply because in [8] he was interested in the question of the definite integral; and if a function assumes infinite values on a set of positive measure, as he wrote, "the integral of such a function cannot be calculated" [8, p. 248]. Similar restrictions naturally appeared at the time in studying other questions, and when it became necessary to discuss convergence in measure, it was measurable almost everywhere finite functions that were meant, even if this was not explicitly stated. In his 1926 book Hobson [3, p. 238] states this requirement from the outset.

The transition to the study of functions assuming infinite values on a set of positive measure and the application of convergence in measure to them required first of all a more precise delineation of the class of functions under consideration and second a change in the definition of convergence in measure itself.

It had been traditional at least since the beginning of the nineteenth century to regard as legitimate a function assuming infinite values of a definite sign at particular points. After the introduction of the principle of negligibility of sets of measure zero it became equally legitimate to consider functions whose infinite values were assumed on a set of measure zero. However, when a function defined almost everywhere was discussed, it was understood that it assumed finite values on the whole set under consideration except possibly at the points of a set of measure zero, where it might become infinite, have finite limits of indeterminacy, or be undefined in general.

Men'shov began to look differently at functions defined almost everywhere. To be specific, he said that "a function is defined almost everywhere on a set if almost everywhere on that set it has a definite value, finite or infinite and of definite sign" [3, p. 6, footnote 2]. And, for example, if $f(x) = +\infty$ almost everywhere on a set of positive measure under consideration, then it is just as regular as a function that is finite almost everywhere. It was in order to study such functions that Men'shov generalized the notion of convergence in measure. He defined the latter as follows:

Suppose given a sequence of measurable functions
$$f_1(x), f_2(x), \ldots, f_n(x), \ldots \qquad (3)$$
finite almost everywhere on some closed interval $[a, b]$. The sequence (3) is said to *converge in measure* on the closed interval $[a, b]$ to the measurable function $f(x)$ defined almost everywhere on $[a, b]$ if almost everywhere on $[a, b]$ the following equalities hold:
$$f_n(x) = g_n(x) + \alpha_n(x) \quad (n = 1, 2, \ldots), \qquad (4)$$
where $g_n(x)$ and $\alpha_n(x)$ are measurable and finite almost everywhere on $[a, b]$ and
$$\lim_{n \to \infty} g_n(x) = f(x)$$
almost everywhere on $[a, b]$, while the sequence $\alpha_n(x)$ $(n = 1, 2, \ldots)$ converges to zero in measure on the given closed interval in the sense defined earlier (p. 123), cf. [3, pp. 3–4].

This is the kind of convergence for which Men'shov proved the following fundamental theorem:

For any measurable function $f(x)$ that is finite almost everywhere on $[0, 2\pi]$ or equal to $+\infty$ or $-\infty$ on a set of positive measure, there exists a trigonometric series converging in measure on $[0, 2\pi]$ to the function $f(x)$, and moreover satisfying the conditions[50]
$$\lim_{n \to \infty} a_n = 0, \quad \lim_{n \to \infty} b_n = 0.$$

---

[50] According to Talalyan [1, p. 79] this result was obtained by Men'shov in 1947 and a preliminary communication of it was given in 1948.

This theorem is a special case of more general results of Men'shov established in the same paper and relating to the limits of indeterminacy in measure of sequences of measurable functions. Since we have not even mentioned the upper and lower limits of sequences of functions, which have a long history going back to Cauchy and Du Bois-Reymond, we shall not give Men'shov's definitions of the concepts of upper and lower limits in measure.

The theorem of Men'shov given here is one of the strongest results in the problem of analytic representation of measurable functions. His paper [3] was the source of many studies by Men'shov himself and a large group of his students and successors; in questions of the representation of functions by sequences of functions convergent in measure the works of Talalyan stand out especially. We cannot go into detail about them,[51] and we limit ourselves to the remark that in 1960 Talalyan [1, pp. 101–108], by modifying the definition of convergence in measure proposed by Men'shov, generalized Men'shov's theorem from the trigonometric system to an arbitrary normalized basis of the space $L^p(G)$, where $p > 1$ and G is a measurable set of positive measure, finding a significantly shorter proof than that of Men'shov.

One may also observe that to represent measurable functions in Men'shov's generality the stronger convergence in the metric of L^p introduced by Talalyan in 1962 suffices [2, p. 369]; in this situation such functions are represented not only by trigonometric series converging in this sense, but also by series over normalized bases of the space L^p [2, p. 370]. These results were generalized by Braun [1] in 1972 to a representation of measurable functions by series in Schauder bases converging asymptotically with respect to the norms of a certain class of Banach spaces containing the space L^p as a special case.

3.7 Convergence in square-mean. Harnack's unsuccessful approach

Mathematicians are not fully in agreement as to the term *convergence in mean*. It was mentioned in the preceding section that Hobson applied the term *convergence on the average* to convergence in measure. In Natanson's book *Theory of Functions of a Real Variable* the "convergence in mean" of a sequence $\{f_n(x)\}$ to a function $f(x)$ is defined by the relation

$$\lim_{n\to\infty} \int_a^b [f_n(x) - f(x)]^2 \, dx = 0 \tag{1}$$

---

[51] We refer the reader to the surveys of Ul'yanov [1; 2, pp. 3–13].

for square-integrable functions $f_n(x)$ and $f(x)$ on (a, b) [1, Vol. I, p. 168]. In Halmos' *Measure Theory* the following definition is given:

> We shall say that a sequence $\{f_n\}$ of integrable functions **converges in the mean,** or **mean converges** to an integrable function f if
>
> $$\rho(f_n, f) = \int |f_n - f|\, d\mu \to 0 \quad \text{as } n \to \infty. \tag{2}$$

[1, p. 103].

Here $\rho(f, g)$ denotes the distance between the functions f and g [1, p. 98]. Besides the fact that essentially different integrals are intended in (1) and (2), these definitions differ in that the square of the difference occurs in the integrand of (1) while the absolute value of the difference occurs in (2). In their book *Elements of the Theory of Functions and Functional Analysis* Kolmogorov and Fomin call convergence of type (1) with a more general integral than Natanson uses *square-mean convergence* [1, Vol. 2, p. 84].

In the present section and the one following the term *convergence in mean* will be understood in the sense of (1) with the Lebesgue integral in its original form.

Convergence in mean, both in the sense of (1) and in the sense of (2), is more general than convergence in measure; historically the former was introduced before the latter. Nevertheless we have chosen to consider them in the opposite order. Our grounds for doing so are that convergence in mean was the source of the concepts of strong and weak convergence, which play such a large role in functional analysis, and in this one of the main epochs in the growth of the theory of functions into functional analysis shows up very clearly.

We shall begin to study the question of the introduction of the concept of convergence in mean at a slightly earlier period than in the preceding section, on the still stronger grounds that, while mathematicians worked implicitly with convergence in measure for a long time—from 1885 to 1909—the attempts to introduce convergence in mean in explicit form, though not completed for objective reasons, was undertaken as early as 1880. It is extremely interesting, and we have ventured to devote an entire section of the present essays to it.

It was noted in the second section of this essay that the mastery of uniform convergence by mathematicians led to a re-examination of many questions of the theory of functions, especially the theory of trigonometric series. This re-examination took place in the 1870's and 1880's, and Harnack joined the effort starting in 1880.[52] The history of the concept of convergence in mean should evidently begin with his memoir [1].

From the outset Harnack takes as the basic premise of his research the assumption that the functions considered are square-integrable[53] [1, p. 124]. He remarks that if $f(x)$ is bounded,[54] the integrability of $[f(x)]^2$ follows from the integrability of $f(x)$; in the case of unbounded functions the situation is different, so that the class of integrable functions is larger than the class of square-integrable functions.

The main theorem of this paper of Harnack's is the following: *If $[f(x)]^2$ is integrable on $(-\pi, \pi)$, then the Fourier series*

$$b_0 + \sum_{k=1}^{\infty} a_k \sin kx + b_k \cos kx \qquad (3)$$

possesses the property that for any arbitrarily small positive number δ there exists a value of n such that not only are all the coefficients a_n and b_n from that value on less than δ in absolute value, but

$$\sum_{k=n}^{n+l}(a_k^2 + b_k^2) < \delta$$

for every natural number l [1, pp. 124–125].[55]

Rephrasing this statement, one can say that if the series (3) is the Fourier series of a square-integrable function, then the series $\sum_{n=1}^{\infty}(a_n^2 + b_n^2)$ converges.[56]

Denoting the remainder of the series (3) by

$$R_{n,l} = \sum_{k=n}^{n+l} a_k \sin kx + b_k \cos kx \quad (l = 1, 2, \ldots),$$

---

[52] For his life and work cf. Gaiduk [1].

[53] In the Riemann sense; throughout this section integrability will be understood in this sense.

[54] Following the custom of the majority of mathematicians of the time, Harnack speaks of the *finiteness* of $f(x)$.

[55] In the statement of the theorem we have basically preserved the terminology and notation of Harnack.

[56] For a proof cf., for example, Tolstov [1, p. 54]; this proof does not differ in essence from Harnack's proof.

3.7 Convergence in square-mean. Harnack's unsuccessful approach

Harnack remarks that the meaning of this theorem, expressed in terms of the remainders of the series (3), is that

$$\int_{-\pi}^{\pi} R_{n,l}^2(x)\,dx < \delta$$

for all l and sufficiently large n. This reminds him of the concept of uniform convergence of the series, where the inequality $|R_{n,l}(x)| < \delta$ is required. He infers from this that this theorem involves a new kind of convergence that generalizes uniform convergence and introduces the concept of "a series uniformly convergent in general" [1, p. 126].

> I shall now call a series *uniformly convergent in general* (im Allgemeinen gleichmässig convergent)[57] if there exists a value n such that for every value of l
>
> $$\int_a^b R_{n,l}^2\,dx < \delta.$$

This is nothing more than convergence in mean for the special case when the integral in (1) is taken in the sense of Riemann. The introduction of this kind of convergence enabled Harnack to reformulate the theorem in the new terminology: *The Fourier series of a square-integrable function converges in mean* [1, p. 126], and he points out in a footnote that it holds not only with respect to the trigonometric series (3) but also for Fourier series in other orthogonal systems.

Since termwise integration of series plays an important role in the theory of trigonometric series, Harnack considers it possible to prove that the Fourier series of a square-integrable function can be integrated termwise [1, pp. 126–127].[58]

In the paper [1] Harnack is primarily interested in the question of analytic representation of an arbitrary square-integrable function. By that time the very meaning of the representation of a function by a Fourier series had changed in comparison with the meaning that was assigned to it when speaking of a representation by a series (1) that is convergent everywhere. Mathematicians had long since neglected finite sets and even certain kinds of infinite sets at which either the series (1) does not convergence or nothing is known about its convergence or divergence. Harnack went a step further in this regard [1, p. 128].

---

[57] We shall refer to such a series as *a series convergent in mean* in what follows.
[58] For a statement and proof of this theorem cf., for example, Tolstov [1, p. 126].

The function $f(x)$ will be said to be *representable* on the interval from a to b using a series if the value $\varphi(x)$ of the series possesses the property that

$$\int_a^b [f(x) - \varphi(x)]^2\, dx = 0.$$

In other words, $f(x)$ is representable using a series if that series converges to $f(x)$ in mean. The latter would have been too good for 1880, if it had not been for one essential circumstance. The function $\varphi(x)$ in Harnack's definition was tacitly assumed to be Riemann-integrable. However, as we now know, the space of functions that are square-integrable in this sense is not complete, so that Harnack's $\varphi(x)$ might fail to be integrable, as Harnack himself soon recognized. He had not noticed this in 1880, and this, together with other things, led him to a series of erroneous conclusions. Nevertheless, besides the intrinsically interesting change in viewpoint on the nature of analytic representation of functions, Harnack's approach played an important role in another respect.

This approach led him to the question of the nature of the set of points at which a mean-convergent series does not converge in the ordinary sense. And Harnack introduced the following class of sets:

A set of points is called a set *of first kind* if its points can be enclosed in intervals of finite length whose total length can be made arbitrarily small [1, p. 128].

Since these sets later played an important role in Harnack's studies and those of other scholars of the nineteenth century who studied the theory of functions, we shall pause to give a more detailed discussion of them.[59]

The first remark involves the term *set of first kind*. Cantor had earlier applied this name to a set whose derived set of some finite order is empty. By 1880 sets of this kind had found many applications in a wide variety of questions. Harnack knew about this, but in formulating this definition he probably did not know that the class of sets he was introducing was larger than the Cantor class of sets of first kind, and for that reason he used Cantor's term. He soon noticed the difference and began calling them *discrete sets*. It was under this name that he subsequently applied them.[60]

---

[59] For this question cf. Medvedev [2, pp. 113–115] and Hawkins [1, pp. 58–61].

[60] In the book of Medvedev [2, p. 113] the introduction of these sets was erroneously assigned to a later paper of Harnack.

3.7 Convergence in square-mean. Harnack's unsuccessful approach

The second remark involves the number of intervals enclosing the set in question. In the definition of Harnack just given nothing is said about this, and if it were possible to consider the number of enclosing intervals countable, we would have here a definition of a set of measure zero in the sense of Borel and Lebesgue. However, the following year Harnack was more precise about this definition, including in it the requirement that the number of enclosing intervals be finite, though allowing it to be arbitrarily large.[61] But even with such a restriction the concept introduced by Harnack marked a significant advance in the evolution of the theory of sets and functions—indeed, his discrete sets are the sets of measure zero in the definition of Peano-Jordan; this definition of measure was put forward by Peano only in 1887 and by Jordan only in 1892.

Harnack wished to use the discrete sets he had introduced [1, p. 131] in order to prove the following theorem:

If $f(x)$ denotes an arbitrary function whose square is integrable on the interval from $-\pi$ to π, its Fourier series represents the value of the function in such a way that the value $\varphi(x)$ of the sum of this Fourier series differs from the value of $f(x)$ by a given amount δ only at the points of a set of first kind [a discrete set—F.M.]. *Moreover*

$$\int_{-\pi}^{\pi} [f(x)]^2 \, dx = 2\pi b_0^2 + \pi \sum_{k=1}^{\infty} (a_k^2 + b_k^2).$$

Although not intrinsically correct, this theorem is interesting for the reasoning by which Harnack tried to prove it. He was unable to find a proof at first, and turned to Du Bois-Reymond. The latter informed him that for this purpose, one could use the following theorem of his own: *If the integral*

$$\int_a^b \lambda(x) g(x) \, dx$$

is zero for any function $\lambda(x)$ having a continuous derivative, then $g(x)$ is an integrable function whose integral over any interval in (a, b) is zero [5, p. 302].

Harnack was dissatisfied with the generality in which Du Bois-Reymond stated the theorem, and he decided to modify it so as to obtain directly the

---

[61] On the other hand Harnack used countable coverings also in 1885; for this cf. Hawkins [1, pp. 63–64].

theorem he had stated. Before considering this proof, we make the following remark. Harnack, as we have said, had proved correctly that

$$\int_{-\pi}^{\pi} R_{n,l}^2 \, dx = \pi \sum_{k=n}^{n+l} (a_k^2 + b_k^2) < \delta, \tag{I}$$

but had erroneously assumed that the sum $\varphi(x)$ of the Fourier series of a square-integrable function is Riemann integrable. Therefore, as he thought, he had succeeded in proving the equalities

$$\int_{-\pi}^{\pi} \varphi(x) \sin kx = \pi a_k = \int_{-\pi}^{\pi} f(x) \sin kx,$$

$$\int_{-\pi}^{\pi} \varphi(x) \cos kx = \pi b_k = \int_{-\pi}^{\pi} f(x) \cos kx \qquad (k = 0, 1, 2, \ldots) \tag{II}$$

[1, pp. 126–127]. Then setting

$$\varphi(x) = f(x) + g(x),$$

he obtains from (II) the equalities

$$\int_{-\pi}^{\pi} g(x) \sin kx \, dx = 0, \quad \int_{-\pi}^{\pi} g(x) \cos kx \, dx = 0 \quad (k = 0, 1, 2, \ldots) \tag{IIa}$$

[1, p. 129]. We now give Harnack's proof in his own words [1, pp. 130–131].

> We construct a function $\lambda(x)$ equal to zero throughout the interval from $-\pi$ to π except on an arbitrarily small interval from α to β, on which it has the constant value 1.
> By Dirichlet's Theorem [on the representability of a function satisfying the famous Dirichlet condition by a trigonometric series—F. M.] such a function can be represented by a Fourier series
>
> $$\lambda(x) = \sum_{k=0} a_k' \sin kx + b_k' \cos kx, \tag{III}$$
>
> in which the coefficients have the value
>
> $$a_k' = \frac{1}{\pi} \int_{-\pi}^{\pi} \lambda(x) \sin kx \, dx, \quad b_k' = \frac{1}{\pi} \int_{-\pi}^{\pi} \lambda(x) \cos kx \, dx.$$

3.7 Convergence in square-mean. Harnack's unsuccessful approach

At points of jump this series has the value 1/2. But, as was shown in the first section, this series converges uniformly in general [in mean—F. M.], and so

$$\int_{-\pi}^{\pi} \lambda(x)g(x)\,dx = \sum_{k=0} \left[a'_k \int_{-\pi}^{\pi} g(x)\sin kx\,dx + b'_k \int_{-\pi}^{\pi} g(x)\cos kx\,dx \right].$$

Since $\lambda(x)$ has the value 1 only on the interval from α to β and is zero at all other points, it follows from the equalities (IIa) that the right-hand side is zero, and hence

$$\int_{-\pi}^{\pi} g(x)\lambda(x)\,dx = \int_{\alpha}^{\beta} g(x)\,dx = 0, \quad -\pi \leq \alpha < \beta \leq +\pi. \quad \text{(IV)}$$

Using this equality, in which the limits of integration α and β can be chosen arbitrarily close together, we obtain the assertion that the points at which the absolute value of $f(x)$ is larger than some value δ form only a set of first kind [a discrete set—F. M.]. Indeed, since the function $g(x)$ is integrable, in any neighborhood of every point α one can define a finite interval in which the oscillation becomes less than δ; if this interval extends from $\alpha + \varepsilon$ to β, then for every point ξ inside the interval

$$\int_{\xi}^{\beta} g(x)\,dx = 0 = (\beta - \xi) \ [g(\xi) + (< \delta)].$$

Consequently the absolute value of $g(x)$ is everywhere less than δ, and the points for which this property does not hold can be enclosed in intervals $\pm\varepsilon$ whose sum becomes arbitrarily small. From equality (I)

$$\varphi(x) = \sum_{k=0} a_k \sin kx + b_k \cos kx = f(x) + g(x)$$

it now follows that

$$\int_{-\pi}^{\pi} \varphi(x)g(x)\,dx = 0 = \int_{-\pi}^{\pi} f(x)g(x)\,dx + \int_{-\pi}^{\pi} g(x)^2\,dx.$$

But since $g(x)$ is in general zero [i.e., zero except on Harnack's discrete set—F. M.], it follows that

$$\int_{-\pi}^{\pi} f(x)g(x)\,dx = 0,$$

Chapter 3 Sequences of functions. Various kinds of convergence

so that we also have

$$\int_{-\pi}^{\pi} [g(x)]^2 \, dx = \int_{-\pi}^{\pi} [\varphi(x) - f(x)]^2 \, dx = 0. \tag{V}$$

Almost all of this is wrong. But we have not given this long quotation in order to demonstrate how erroneously even good mathematicians can reason. What is much more important is the number of interesting ideas contained in this reasoning.

The grandeur of Harnack's vision is striking. In 1880 he intended no less than to solve a problem that subsequently resisted the minds of Lebesgue, Hilbert, Riesz, and many others at the beginning of the twentieth century: to introduce the space L^2 of square-integrable functions and by solving the integral equations (IIa) to prove that it is complete. Naturally he did not succeed in doing this. But such a failure is worth many results that are irreproachable from the formal point of view. In the arguments just given, besides the ideas of convergence in mean and representability of a function in mean, we find the system of integral equations whose study caused so much trouble afterwards and led in particular to the concepts of strong and weak convergence of sequences of functions;[62] here the principle of the negligibility of sets of measure zero, in an insufficiently broad interpretation, to be sure, is applied as it was later to be applied in the development of the theory of functions; sets of the form $E(|f(x)| > \delta)$ are introduced, which were widely used later in many arguments; Harnack deduced immediately from Eq. (V) that the closure formula holds for every square-integrable function $f(x)$[63] [1, p. 131], etc.

Besides the erroneous equalities (II) and the inferences obtained from them, as well as the theorem just discussed, Harnack attempted in this paper to obtain yet another incorrect proposition to the effect that the uniqueness theorem for representation by trigonometric series remains valid when sets of measure zero in the sense of Peano-Jordan are neglected (Harnack [1, pp. 131–132]). Again the grandeur of his reach is striking: he is trying to solve a problem that continues to challenge mathematicians even a hundred years later.

---

[62] For this cf. Sec. 9 of the present essay.

[63] The closure formula is sometimes called Lyapunov's formula (Tolstov [1, p. 119]). But in fact Lyapunov's proof was never published, and it is known only from communications of Steklov, for example [1, p. 784], and from the title of Lyapunov's report, given in the 1896 *Bulletin of the Khar'kov Mathematical Society*. Many mathematicians studied this question, cf. Paplauskas [1, pp. 184–194].

It is not without interest that although Harnack's vision as a whole was premature, nevertheless other scholars were arriving at similar ideas. In 1882 Hugoniot, almost certainly not knowing of Harnack's paper, established [1] (in modern language) the mean convergence of an expansion of a square-integrable function in any orthonormal system,[64] and in doing so committed a similar error, consisting of the assumption that mean convergence implies ordinary convergence. Halphen noticed this error immediately and constructed an example of a series that converges to zero in mean, but diverges in the ordinary sense everywhere except at one point [1, pp. 1218–1219].

No doubt other mathematicians besides Hugoniot and Halphen used convergence in mean. It seems, however that no one except Harnack saw it as a specific kind of convergence. Hugoniot and Halphen simply asserted the fact that the remainder of the series tends to zero in mean, and it is this that enables us to speak of convergence in mean in connection with their works.

Harnack himself soon realized the weakness in his reasoning in [1]. In an 1882 paper [2], returning to certain propositions of his earlier paper, he abandons convergence in mean and replaces it by uniform convergence almost everywhere [2, p. 251], where the words *almost everywhere* that we have used refer to sets of measure zero in the sense of Peano-Jordan (discrete sets in the terminology of Harnack); the representation of a function in mean is correspondingly replaced by the representation using a series convergent in this sense. He again committed many serious errors. We shall not dwell on them, since Harnack himself recognized them immediately [3], and incidentally discovered one of the sources of the error in his conclusions: the assumption that the integrability of the remainder $R_{n,l}(x)$ for some n and every natural number l implies that the limiting function is integrable. This destroyed his basic conception of the completeness of the Riemann square-integrable functions and evidently spurred him on to his subsequent investigations on generalization of the concept of integral, in which he achieved certain successes.[65]

3.8 Square-mean convergence. The work of Fischer and certain related investigations

Original though Harnack's projects of 1880 were, they were sufficiently in advance of their time that he himself rejected them; in the nineteenth century the soil had not been prepared for their realization. The efforts of a large number of scholars were needed to make it both possible and necessary to

---

[64] Harnack, as mentioned, proved this for the trigonometric system, but remarked that the proposition holds for other orthogonal systems.

[65] For this cf. Pesin [1, pp. 29–32].

return to Harnack's ideas on a higher level of mathematical development. In this connection it is curious that Harnack's ideas were thoroughly forgotten[66] and, as it were, had to be born again. Their birth was connected with the foundation of two mathematical disciplines—the theory of functions of a real variable in its renewed form and functional analysis. For that reason we shall precede our study of the introduction and early applications of mean convergence with two bits of historical information.

To solve the many problems that arose in the theory of functions in the nineteenth century, an essential generalization of the Riemann integral was needed, and such a generalization had appeared at the very beginning of the twentieth century in the form of the Lebesgue integral. Lebesgue [1] first published the new definition of the integral in 1901. He gave a detailed exposition of the new theory of integration and a variety of applications of it in the following year [3]. A year later he proposed two new and important applications of his integral—to prove the existence of a derivative almost everywhere for any monotone continuous function [4] and to generalize the Fourier series to the Fourier-Lebesgue series, along with certain theorems of the theory of trigonometric series [6]. With the appearance of his books [8] and [12] there began a triumphal procession of the new concept of integral in analysis in the broad sense.[67]

It was around this time that the Hilbert-Fréchet functional analysis arose. Hilbert began to publish his investigations in the theory of integral equations in 1904 and five of his papers on this topic had been printed by 1906.[68] Fréchet began to construct abstract functional analysis at the same time; his first papers also date from 1904, and he gave a detailed exposition of his ideas [4] in 1906.[69]

At the time Hilbert was apparently not aware of the theory of the Lebesgue integral and conducted his investigations within the framework of the Riemann integral. Fréchet, in contrast, had already mastered the new concept of integral by 1904, but his thoughts were directed more toward the foundation of abstract conceptions than to specific problems, in particular to the theory of integral equations. In the meantime the more acute mathematical problem arose of generalizing Hilbert's results and putting them in a more definitive analytic form. Many scholars sought a solution to this problem—

---

[66] The only mention of Harnack known to me in this connection is in the survey of Rosenthal [1, p. 1181, footnote 1068].

[67] For more details see the book of Hawkins [1].

[68] The sixth was published in 1910; in 1912 all these works were published as a separate book: Hilbert [1].

[69] For more details see Bernkopf [1].

Riesz, Fischer, Schmidt, and others, who occupied a position in some sense intermediate between Hilbert and Fréchet.

As noted in the preceding section, it was known in the last century that every Fourier series of a function $f(x)$ whose square is Riemann integrable converges in mean, from which it followed that the sum of the squares of its Fourier-Riemann coefficients c_n in an orthonormal system satisfy the inequality

$$\sum_{n=1}^{\infty} c_n^2 < \infty.$$

This result carried over easily to Fourier-Lebesgue series. But at the beginning of the century, principally in connection with the Hilbert theory of integral equations, the converse question became more important: suppose given a sequence of constants $\{c_i\}$ the sum of whose squares converges; does there exist a square-integrable function $f(x)$ whose Fourier-Lebesgue coefficients in a given orthonormal system are the numbers of the sequence $\{c_i\}$? In other words, if we denote by l^2 the space of sequences the sum of whose squares converges and by L^2 the space of square-integrable functions, can one establish an isomorphism between l^2 and L^2? The answer to this question opened the way to the study of many problems of the theory of integral equations and the theory of functions. As we have seen in the previous section, within the framework of the theory of Riemann integration the answer to this question is negative, as Harnack himself had essentially recognized in 1882. The creation of the theory of Lebesgue integration, in contrast, made it possible to give a positive answer, at which both Riesz and Fischer arrived independently in 1907.

Riesz published his results connected with the so-called Riesz-Fischer theorem in two notes [3, 4], of which the second was a summary of the first. Although the reasoning followed by Riesz is not directly connected with the topic of the present section, we shall nevertheless pause briefly to discuss it, since we shall need it in our study of strong and weak convergence.

Hilbert [1] had proposed a general method of studying the linear integral equation

$$\varphi(s) = f(s) + \int_a^b K(s,t) f(t)\, dt \tag{1}$$

for continuous φ and K. His method consisted of noting that solving Eq. (1) reduced to solving a system of an infinite number of linear equations in an infinite number of unknowns. The transition from the integral equation (1) to the system of linear equations is carried out using a certain orthonormal system, and the coefficients and unknowns of the system of linear equations occurred as Fourier coefficients with respect to this orthonormal system.

One of the fundamental questions for Hilbert's method was the following. Suppose to each function defined on $[a,b]$ in the orthonormal system $\{\varphi_i(x)\}$ there is assigned a certain real number a_i. What conditions must the system of numbers $\{a_i\}$ satisfy in order for the equalities

$$\int_a^b f(x)\varphi_i(x)\,dx = a_i$$

to hold? In other words what are the necessary and sufficient conditions for the numbers a_i to be regarded as the Fourier coefficients of an unknown function $f(x)$ in the system $\{\varphi_i(x)\}$? Under the assumption that $\varphi(s)$ and $K(s,t)$ are continuous Hilbert gave the answer, appealing to the theory of quadratic forms.

Riesz, wishing to generalize Hilbert's results, arrived at a theorem that reads as follows in his formulation [3, p. 390]: *Let $\{\varphi(x)\}$ be a normalized orthogonal system of square-integrable* [in the Lebesgue sense—F. M.] *functions defined on the interval ab, i.e., a system for which*

$$\int_a^b \varphi_i(x)\varphi_j(x)\,dx = 0 \ (i \neq j) \ \text{and} \ \int_a^b [\varphi_i(x)]^2\,dx = c^2$$

for every function of the system. We now assign to each function of this system a number a_i. Then the convergence of $\sum a_i^2$ is both a necessary and a sufficient condition for the existence of a square-integrable function $f(x)$ such that

$$\int_a^b f(x)\varphi_i(x)\,dx = a_i.$$

This is the famous Riesz-Fischer Theorem. Riesz' proof of this theorem is rather complicated and required an appeal to strong results of the theory of functions, such as Lebesgue's theorem that a function of bounded variation has a derivative almost everywhere and the Poisson summation method.

The route by which Fischer arrived at the same theorem is less clear. The reason is that he already knew about the note of Riesz [4] when he sent his article [1] for publication, although he had arrived at his results before the publication of Riesz' results, as he stated in print [1]. Evidently the source was the same—Hilbert's theory, only considered from a slightly different point of view. But since Riesz had pointed out this source in [4], there was no need for Fischer to do so.

Fischer's method of proving this theorem is in general simpler. The simplification was achieved by introducing the concept of convergence in mean of a sequence of functions.

Fischer called a sequence $\{f_n(x)\}$ of square-integrable functions *convergent in mean* (en moyenne) if

$$\lim_{m,n \to \infty} \int_a^b [f_n(x) - f_m(x)]^2 \, dx = 0. \qquad (2)$$

He said that this sequence converges in mean to a function $f(x) \in L^2$ if[70]

$$\lim_{n \to \infty} \int_a^b [f(x) - f_n(x)]^2 \, dx = 0. \qquad (3)$$

The main theorem of Fischer's article is the following: *If a sequence of functions belonging to Ω converges in mean, then there exists a function f in Ω to which it converges in mean* [1, p. 1023]. In other words, the space L^2 is a complete space with respect to convergence in mean.[71] From this Fischer obtained the Riesz-Fischer Theorem as a corollary.[72] The kind of convergence of sequences of functions introduced by Fischer turned out to be very fruitful and almost immediately found numerous applications. Fischer himself, in another note that followed [1], gave such applications.

Let $\{\varphi_n\}$ be an orthonormal system defined on $[a,b]$. He called functions $\varphi \in L^2[a,b]$ *accessible* using the functions φ_n if, for any $\varepsilon > 0$ there exists a natural number n and constants $c_1, c_2, \ldots c_n$ such that

$$\int_a^b [\varphi - (c_1\varphi_1 + c_2\varphi_2 + \cdots + c_n\varphi_n)]^2 \, dx < \varepsilon.$$

Relying on this definition, he proved that for a given function $f \in L^2[a,b]$ there exists one and only one function φ accessible using the functions φ_n for which the integral

$$\int_a^b (f - \varphi)^2 \, dx$$

is a minimum. He called it the accessible function giving the best mean approximation to the function $f(x)$ [2, pp. 1148–1149].

In terms of convergence in mean Fischer established here necessary and sufficient conditions for a continuous function to be the indefinite integral of a square-integrable function [2, pp. 1149–1151].

---

[70] Fischer did not use the notation $f \in L^2$. He spoke of the set Ω of square-integrable functions.

[71] Cf. Natanson [1, Vol. I, pp. 170–171].

[72] Cf. Natanson, [1, Vol. I, pp. 179–180].

In 1909–1910 Riesz generalized Fischer's square-mean convergence; we shall return to this subject in the next section. Just now we shall continue our survey of the studies and applications of convergence of this form. Weyl [1, pp. 243–244] established in 1909 that from every mean-convergent sequence of functions one can extract a subsequence that is convergent almost everywhere[73] to the same limit and also gave another proof [1, pp. 244–245] of the Riesz-Fischer Theorem. These last results of Weyl were obtained anew by Plancherel [1, pp. 292–296] in the following year in a slightly modified form. It is curious that in doing so Plancherel, like Weyl, did not use the terminology of convergence almost everywhere, but followed Weyl in saying that a sequence converges uniformly in general (uniformément en général). In this paper Plancherel remarked that a mean-convergent sequence can be integrated termwise [1, p. 297]. In 1913 Radon generalized mean convergence even further than Riesz, as will also be discussed in the next section. In this same year Pincherle proposed a peculiar modification of mean convergence. Unfortunately we do not have his work at our disposal,[74] and the following discussion is based on a summary of these results in the memoir of Nalli [3, pp. 149–150].

In Fischer's definition the sequence of functions $\{f_n(x)\}$ converging in mean to $f(x)$ is assumed countable. This restriction was natural, since one of the principal motivations for introducing this kind of convergence was the intention of studying the problem of representing function by a series of orthogonal functions, in particular, to obtain the Riesz-Fischer Theorem. But, as Schmidt [1] and Riesz [1] had shown in 1906 (the former for continuous functions, the latter for Lebesgue-integrable functions), every complete orthonormal system is countable. However, in a variety of questions it is necessary to consider uncountable systems of orthogonal functions also; it was for that case that Pincherle extended the concept of mean convergence.

A system of real-valued functions $f(t, r)$, defined on $[a, b]$ and depending on a real parameter r, each of which is square-integrable, is said to be convergent in mean to a square-integrable function $f(t)$ as $r \to r_0$ if for every $\varepsilon > 0$ there exists $\delta > 0$ such that $|r - r_0| < \delta$ implies

$$\int_a^b [f(t) - f(t, r)]^2 \, dt < \varepsilon.$$

Using this kind of mean convergence Pincherle found more general conditions than had previously been known for a function of a complex variable to be represented by a Cauchy integral.

---

[73] Uniformly convergent in general, according to Weyl.

[74] S. Pincherle, *Un'applicazione della convergenza in media*, Rend. della R. Acc. dei Lincei, **5**, XXII, 2 sem., 1913, pp. 397–402.

We shall now pause to discuss the generalization of Nalli.

To make this generalization more transparent, we recall that Fischer's definitions (2) and (3) imply that for every mean-convergent sequence of functions $\{f_n(x)\}$ there exists a function $f(x) \in L^2[a,b]$ such that

$$\int_a^b [f(x)]^2\, dx = \lim_{n\to\infty} \int_a^b [f_n(x)]^2\, dx. \tag{4}$$

Here the function $f(x)$ is not unique, but any two such functions are equivalent, i.e., if $\{f_n(x)\}$ also converges in mean to $h(x)$, then

$$\int_a^b [f(x) - h(x)]^2\, dx = 0. \tag{5}$$

In particular if

$$f_n(x) = \sum_{j=1}^n a_j \varphi_j(x), \tag{6}$$

where $\{\varphi_j(x)\}$ is a sequence of orthonormal functions and $\{a_n\}$ is a sequence of constants with $\sum_{n=1}^{\infty} a_n^2 < \infty$, then there exists a function $f(x) \in L^2[a,b]$ for which

$$\int_a^b f(x)\varphi_n(x)\, dx = a_n, \quad \int_a^b [f(x)]^2\, dx = \sum_{n=1}^{\infty} [a_n]^2,$$

and $f(x)$ is again unique up to equivalence (the Riesz-Fischer Theorem).

Nalli's first step was to restate all these results not only for functions of a real variable, but also for functions of a complex variable [1, pp. 305–306]. Such a restatement is quite obvious; one need only replace the brackets in Eqs. (4)–(7) by absolute value signs and write the orthonormality condition in the form

$$\int_a^b \varphi_i(x) \bar\varphi_j(x)\, dx = \begin{cases} 0 & (i \neq j), \\ 1 & (i = j), \end{cases}$$

where $\bar\varphi_j(x)$ is the complex conjugate of the function $\varphi_j(x)$.

Nalli's second step was more significant. She posed the following problem.

Suppose given a sequence $\{f_n(s)\}$ of functions whose absolute values are square-integrable on every finite interval and which satisfy the conditions

$$\lim_{m,n\to\infty} \left[\lim_{\omega\to\infty} \frac{1}{2\omega} \int_{-\omega}^{\omega} |f_m(s) - f_n(s)|^2\, ds \right] = 0,$$

$$\lim_{\omega\to\infty} \frac{1}{2\omega} \int_{-\omega}^{\omega} |f_n(s)|^2\, ds = l_n, \tag{2'}$$

where $\{l_n\}$ is a sequence of numbers. It is necessary to construct a function $g(s)$ for which the equalities

$$\lim_{n\to\infty}\left[\lim_{\omega\to\infty}\frac{1}{2\omega}\int_{-\omega}^{\omega}|g(s)-f_n(s)|^2\,ds\right]=0 \qquad (3')$$

and

$$\lim_{\omega\to\infty}\frac{1}{2\omega}\int_{-\omega}^{\omega}|g(s)|^2\,ds=\lim_{n\to\infty}\left[\lim_{\omega\to\infty}\frac{1}{2\omega}\int_{-\omega}^{\omega}|f_n(s)|^2\,dx\right] \qquad (4')$$

hold, then obtain from the solution a generalization of the Riesz-Fischer Theorem to an infinite interval, and not only for countable sequences of functions.

It is clear that conditions (2') are the definition of a mean-convergent sequence of functions whose absolute values are square-integrable on an infinite interval; the function $g(s)$ in (3') plays the role of the function $f(x)$ in (3), and Eq. (4') is the corresponding analogue of Eq. (4).

Nalli [1] solved this problem under the following assumptions. Suppose

$$f_1(s), f_2(s), \ldots, f_n(s), \ldots$$

is a sequence of functions defined on the infinite interval $(-\infty, +\infty)$, uniformly bounded, and satisfying the following conditions:

I. $\quad \eta_{m,n} = \lim_{\omega\to\infty}\frac{1}{2\omega}\int_{-\omega}^{\omega}|f_m(s)-f_n(s)|^2\,ds$

exists for all m and n.

II. $\quad \lim_{m,n\to\infty}\eta_{m,n}=0.$

III. The following limit exists:

$$\lim_{\omega\to\infty}\frac{1}{2\omega}\int_{-\omega}^{\omega}|f_n(s)|^2\,ds$$

for any n. Then there exists a function $g(s)$ defined at all points of the interval $(-\infty, +\infty)$ and bounded and Lebesgue integrable along with the square of its absolute value on every finite interval for which Eqs. (3') and (4') hold.

The Riesz-Fischer Theorem for an infinite interval follows from this theorem [1, pp. 318–319].

In another paper [2] Nalli eliminated the requirement that the sequence $\{f_n(s)\}$ be uniformly bounded, requiring only that it be bounded, and assuming that the numerical sequence $\{l_n\}$ is increasing or at least contains an increasing subsequence.

In 1916, wishing to apply the kind of convergence she had introduced in the theory of Dirichlet series, and at the same time to generalize the results of Pincherle, Nalli proposed another modification of mean convergence [3, p. 152].

Let $\{f(t,r)\}$ be a system of real or complex functions of a real variable t depending on the parameter r. Let $f(t,r)$ be square-integrable in absolute value on each finite interval for fixed r, and in addition assume that

$$\limsup_{\omega \to \infty} \frac{1}{2\omega} \int_{-\omega}^{\omega} |f(t,r)|^2 \, dt = L_r,$$

where the numbers L_r are finite. The set of functions $\{f(t,r)\}$ is said to be *convergent in mean* as $r \to r_0$ on the interval $(-\infty, +\infty)$ to a function $f(t)$ whose absolute value is square-integrable on every finite interval if for every fixed $\varepsilon > 0$ there exists $\delta > 0$ such that for $|r - r_0| < \delta$

$$\limsup_{\omega \to \infty} \frac{1}{2\omega} \int_{-\omega}^{\omega} |f(t) - f(t,r)|^2 \, dt \leq \varepsilon \quad [3, \text{p. } 152].$$

We shall not go into the details of Nalli's applications of this convergence.

The concept of convergence in mean of a sequence of functions was applied explicitly or implicitly in many other questions, in particular in the question of the closure of orthonormal systems, for example in the works of Steklov,[75] Severini [1, 2], and Privalov [1]. We note that in both of his papers Severini actually used a kind of mean convergence in which the integrand in (2) and (3) contains a weight function as a factor; and Privalov, in particular, proved that for any bounded measurable function there exists a sequence of continuous functions converging to it in mean, while for every square-integrable function there exists a sequence of bounded measurable functions converging to it in mean.

3.9 Strong and weak convergence

Riesz' paper "Untersuchungen über Systeme integrierbarer Funktionen" [7] is the classic work from which many of the ideas and methods of the theory of functions and functional analysis arose. Our intention is not by any means to give a comprehensive description of its content.[76] Our intention is much more modest: to discuss certain reasoning of Riesz that led him to the generalizations of convergence of sequences whose names form the title of this

---

[75] For this cf. Paplauskas [1, pp. 187, 190–194].

[76] To a significant degree, although not completely, this has been done by Bernkopf [1, pp. 54–62].

section and to point out certain questions connected with this. It is in this paper also that Riesz introduced the generalization of convergence in square mean mentioned in the preceding section.

His starting positions here are the same as described at the beginning of the preceding section. As before he starts with a system of integral equations

$$\int_a^b f(x)\varphi_i(x)\,dx = a_i, \tag{1}$$

but now he seeks the solution $f(x)$ not in the space $L^2[a,b]$ but in the space $L^p[a,b]$,[77] defined by him as the collection of functions $f(x)$ such that the integral

$$\int_a^b |f(x)|^p\,dx, \quad p > 1,$$

exists [7, p. 451].[78]

We have already said (p. 140) that for $L^2[a,b]$ Riesz obtained the solution in the form of the Riesz-Fischer Theorem in a rather complicated way. Since Fischer obtained the same theorem in a significantly simpler way, Riesz had evidently now decided to follow his route. He had pointed out this route in his 1909 paper [5], in which the concept of convergence of a sequence in measure was introduced. Indeed, in the conclusion of this note, he formulated the following theorem [5, p. 399]: *Let p be some positive number and $|f_n(x)|$ a sequence of measurable functions such that the integrals*

$$I_{i,k} = \int_E |f_i(x) - f_k(x)|^p\,dx$$

exist and

$$\lim_{i=\infty,\,k=\infty} I_{i,k} = 0.$$

Then there always exists a measurable function $f(x)$ such that

$$\lim_{n=\infty} \int_E |f(x) - f_n(x)|^p\,dx = 0.$$

It is clear that these words contain a generalization of the Fischer definition of convergence in mean of a sequence of functions; and although Riesz

---

[77] Riesz spoke of the functional class $[L^p]$.
[78] It is the existence of a finite integral that is meant; Riesz did not specify this.

did not say so directly at this point, it follows from his words that this proposition contains the basic theorem of Fischer on the completeness of L^2 as a special case.

It would seem that the way was now open for Riesz to find the desired solution of the system (1): a new class of functions had been introduced in which he wished to seek the solution, along with the corresponding kind of convergence; one has only to prove the analogue of the Riesz-Fischer theorem for this case, and the problem will have been solved. Nevertheless he encountered a certain difficulty that had not arisen in studying the space L^2, i.e., when $p = 2$.

This difficulty consisted of the fact that while the functions of the given orthonormal system $\{\varphi_i(x)\}$ and the desired function $f(x)$ could be taken in the same space L^2 when solving the system (1) for $p = 2$, this turned out not to be the case for $p \neq 2$. The function-theoretic reason for this is that the product of two integrable functions is not always integrable, and one of the crucial points in Riesz' paper [7] is the discovery of conditions under which the product is integrable.

Using the Hölder inequality[79]

$$\left| \int_E f(x)\varphi(x)\,dx \right| \leq \left[\int_E |f(x)|^p\,dx \right]^{\frac{1}{p}} \left[\int_E |\varphi(x)|^{\frac{p}{p-1}}\,dx \right]^{\frac{p-1}{p}}, \qquad (2)$$

he easily obtains a sufficient condition: the product of two integrable functions $f(x)$ and $\varphi(x)$ is integrable if there exists a number $p > 1$ such that the functions $|f(x)|^p$ and $|\varphi(x)|^{\frac{p}{p-1}}$ are integrable [7, p. 449]. This condition is also necessary in a certain sense. To be specific Riesz proved [7, pp. 449–451] that if the product $f(x)\varphi(x)$ is integrable for all functions $f(x)$ whose absolute values raised to power p are integrable, then the function $|\varphi(x)|^{\frac{p}{p-1}}$ is integrable.[80]

Since the solution of the fundamental problem posed by Riesz required precisely the study of the integral of the product of integrable functions (the

---

[79] We have preserved Riesz' notation from [7, p. 448]. Here he did not explicitly introduce the exponent q conjugate to p and satisfying the equality $\frac{1}{p} + \frac{1}{q} = 1$; for him the role of q was played by the number $\frac{p}{p-1}$.

[80] It is probably worthwhile to point out that Lebesgue had studied this question a year earlier than Riesz and obtained many sufficient conditions for integrability of a product of functions [14, pp. 34–39], the most general of which was that the product of an integrable function and a function equivalent to a bounded function is integrable. Riesz, who was more interested in the problems of functional analysis, managed to find a more beautiful condition and, more importantly, one that was more useful at the time.

coefficients of series expansions in a given orthonormal system are expressed by such integrals), the propositions he had established suggested that together with the space L^p one needed to introduce another functional space—the space of functions for which the $p/(p-1)$ power of the absolute value is integrable. Having introduced it, Riesz obtains the following proposition from the latter theorem and Hölder's inequality: *Each of the spaces L^p and $L^{p/(p-1)}$ consists precisely of the functions that yield an integrable product when multiplied by any function of the other space* [7, p. 451]. If in particular $p = 2$, then $p/(p-1) = 2$ also and so a function that forms an integrable product when multiplied by every function of L^2 is itself a member of L^2.

> This property of the class $[L^2]$ is the basis for the important role it plays in all questions involving the integral of a product. If we now attempt to extend the results that hold for the class $[L^2]$ to the other classes $[L^p]$, we must consider simultaneously the classes $[L^p]$ and $[L^{p/(p-1)}]$ [7, p. 451].

In other words, since L^2 is a Hilbert space and L^p is not a Hilbert space when $p \neq 2$, in studying questions that require the use of an inner product it is necessary to invoke, along with L^p, the space L^q with the exponent q conjugate to p. And Riesz now seeks a solution of the system (1) with $f(x) \in L^p[a,b]$ and $\varphi_i(x) \in L^{p/(p-1)}[a,b]$. Not only that, in order to obtain necessary and sufficient conditions for the solvability of this system, even under this last assumption, it was necessary to apply not only a direct generalization of Fischer's convergence in square mean but also a more general type of convergence. We now turn to consider these forms of convergence.

Riesz called the generalized square-mean convergence *strong convergence with exponent p* and defined it as follows.

Suppose a sequence $\{f_n(x)\}$ and a function $f(x)$ are defined on $[a,b]$ and belong to $L^p[a,b]$. The sequence *converges strongly* to $f(x)$ with exponent p if the equality

$$\lim_{n \to \infty} \int_a^b |f(x) - f_n(x)|^p \, dx = 0$$

holds [7, p. 456].[81] For this equality Riesz established the analogue of Fischer's theorem on the completeness of L^2, i.e., the proposition that if a sequence $\{f_n(x)\}$, with $f_n(x) \in L^p[a,b]$ satisfies the condition

$$\lim_{\substack{i \to \infty \\ j \to \infty}} \int_a^b |f_i(x) - f_j(x)|^p \, dx = 0,$$

---

[81] In the book of Natanson [1, Vol. I, p. 199] this convergence is called *mean convergence of order p*.

3.9 Strong and weak convergence

then there exists a function $f(x) \in L^p[a,b]$ to which this sequence converges strongly with exponent p [7, p. 460]. This proposition, as we have remarked above, had been published a year before this.

However, this kind of convergence was insufficient for solving the basic problem, since the space L^p, $p \neq 2$, is not a Hilbert space; and in stating the preceding theorem, Riesz remarks immediately that he will not use it in [7]. He needed to generalize strong convergence in another direction and study it in more detail in order to achieve the goal he had set for himself. Before discussing this generalization, we make a digression.

It was remarked above (p. 136) that the germ of the idea of strong convergence was contained in Harnack's arguments. It had probably been used in one form or another by many mathematicians of the nineteenth and early twentieth centuries, but only implicitly. Even Lebesgue, who had, so to speak, come closer than anyone else to the idea of weak convergence a year before the appearance of Riesz' paper [7], had not yet distinguished it as an independent concept. We shall pause to discuss his reasoning in more detail, since Riesz referred directly to it [7, p. 458, footnote].

Lebesgue [14, pp. 51–63] started from a functional

$$I_n(f) = \int_a^b f(\alpha)\varphi(\alpha, n)\, d\alpha \qquad (3)$$

defined on a family of functions $f(x)$[82] and sought conditions that the kernel $\varphi(\alpha, n)$ must satisfy in order for the functional to tend to zero as $n \to \infty$. It is obvious that the system of equations (3) is an analogue of the system (1) studied by Riesz, but the goals set by Riesz and Lebesgue were somewhat different. We do not intend to linger over the contents of Lebesgue's paper [14], since at this point we are interested only in the question of the introduction of weak convergence; we merely remark that Lebesgue approached this question armed with the conditions for the integrability of a product mentioned in footnote 80 above. We shall also give one of the results he established: *In order for $I_n(f)$ to tend to zero along with $1/n$ for every square-integrable function f, the following conditions are necessary and sufficient:*

1. *The kernel $\varphi(\alpha, n)$ is square integrable and the integral*

$$\int_a^b \varphi^2(\alpha, n)\, d\alpha$$

---

[82] At the time it was not yet customary to specify precisely the form of the function space in question, and even the term *function space*, though it had been introduced in the nineteenth century by Pincherle, was not precisely defined and was not in common use.

is bounded by a number independent of n;

2. The integral $\int_\lambda^\mu \varphi(\alpha, n) \, d\alpha$ tends to zero along with $1/n$ for any λ and μ in (a, b) [14, p. 55].

In essence the passage just quoted is a explicit statement of weak convergence in L^2, only it is masked by the form Lebesgue gave it. There are several similar theorems in this paper of Lebesgue's, but all are stated in a form similar to this one.

Riesz was already convinced of the usefulness of introducing different kinds of convergence, and for that reason he explicitly introduced weak convergence [7, p. 457]:

> A sequence of functions $\{f_i(x)\}$ of class $[L^p]$ *converges weakly* with exponent p to a function $f(x)$ of the same class if
>
> a) all the values of the integrals
>
> $$\int_a^b |f_i(x)|^p \, dx$$
>
> are less than a certain finite bound;
>
> b) for all points $a \leq x \leq b$ the equality
>
> $$\lim_{i=\infty} \int_a^x f_i(x) \, dx = \int_a^x f(x) \, dx$$
>
> holds.

Riesz shows that strong convergence in L^p implies weak convergence, but that the converse is not true [7, p. 457], introduces the concept of weak compactness in L^p [7, p. 459], and uses it to find necessary and sufficient conditions for the solvability of the system of integral equations (1) [7, pp. 461–466].[83] He applies all this in the remainder of the paper to study linear functionals and operators on L^p.

Riesz worked out all his reasoning in [7] in the context of the usual Lebesgue integral. But by that time he had come to know of the concept of Stieltjes integration in the framework of the Riemann-Stieltjes integral. For that reason a year later he again returned to the system (1) in order to study it under the assumption that the integrals in it are understood in the sense of Stieltjes (Riesz [8]). It would seem that, equipped with the new concept of integral, it would be possible to follow the route laid down in the preceding

---

[83] For a later treatment see the book of Kaczmarz and Steinhaus [1, pp. 227–231].

3.9 Strong and weak convergence

paper [7]. But then Riesz would have found himself in a situation similar to that of Harnack in 1880—faced with the insufficiency of the Riemann-Stieltjes integral. As if sensing this danger, Riesz occupied himself in [8] mainly with other questions connected with the generalization of the system (1), and not with those that had interested him a year earlier. Further progress in the direction laid down in [7] required a new generalization of the concept of integral.

Radon [1], introducing the concept of the Lebesgue-Stieltjes integral in 1913, generalized the basic results of Riesz' paper [7] for this case [1, pp. 1351–1438], constructing a theory parallel to that of Riesz and including the latter as a special case. We shall not give an exposition of Radon's theory, but limit ourselves to what has been said and turn to another extension of the concept of strong convergence that was widely used in studies of the theory of functions and functional analysis.

W. H. Young arrived at the functional spaces L^p independently of Riesz, but he went further. Before discussing this, we permit ourselves a digression on the mathematical works of this scholar.[84]

Young was a very idiosyncratic investigator. Undertaking investigations in the area of mathematics relatively late, at the very beginning he independently obtained many important, but mostly known results. In doing this he arrived at these results along nontraditional paths, developing new methods, and introducing his own mathematical language. These methods (like the method of monotone sequences) enabled him later to obtain many new results.

His mathematical legacy is enormous. Hardy [7, p. 223] pointed out over two hundred papers and three books, to which one must add that the majority of Young's articles are of a size significantly larger than the average mathematical publication, and reading his works is very difficult. The difficulty is partly due to his nontraditional language and partly because few of his works can be satisfactorily understood in isolation from his other studies, but mainly because they are characterized, to use Hardy's phrase, "by an enormous energy and wealth of original ideas," and because Young "possessed a superabundance of ideas, too many for anyone to develop fully" [7, p. 223]. They were developed by Young himself and by his wife and son and daughter; and many others studied these ideas and evidently will do so again many times. One of them—probably not the most significant—was Young's idea of supersummability.

We shall not undertake to trace the genesis of this idea in Young's work,

---

[84] On the life and teaching activity of W. H. Young and his wife G. C. Young, also a well-known mathematician, cf. Grattan-Guinness [3].

and the preceding paragraph was partly intended to justify this refusal. We shall begin with his 1912 paper [4], in which the idea of this "supersummability" is rather distinctly introduced.

In connection with certain problems of the theory of Fourier series and methods of summing them Young was interested in the problem of finding conditions for a given continuous function to be the indefinite integral of an integrable function. In 1904 Lebesgue [8, p. 128] and especially Vitali [1] had given the necessary and sufficient conditions that must be imposed on a given $F(x)$ in order for it to be the indefinite integral of some integrable function $f(x)$—this is the condition that the function $F(x)$ be absolutely continuous. However, in the problems Young was studying something else was needed: he was interested in the conditions under which $F(x)$ is the indefinite integral of an integrable function $f(x)$ satisfying additional conditions that in particular cases reduced to the requirements $f(x) \in L^2$ or $f(x) \in L^p$. As we have seen, Fischer had solved this problem for L^2, and Riesz [7, pp. 454–455] had generalized his solution to L^p.

Young's goal of obtaining an analogous theorem for class of functions for which not just the pth power of the absolute value of $f(x)$ but some general function of $f(x)$ is integrable (hence Young's "supersummability") required the introduction of an important inequality for its realization, one that did not follow from the classical inequalities of Hölder or Minkowski, which Riesz had used in studying the space L^p. Young [4, p. 226] introduced this inequality in 1912, and it enabled him to introduce classes of integrable functions that generalized the spaces L^p and eventually grew into the Orlicz spaces.

In Young's formulation this inequality is as follows: *If $v = U(u)$ is a positive increasing function of a positive real variable u and has a derivative $U'(u)$ that is positive (> 0), while $V(v)$ is the function inverse to it, so that*

$$u = V(v),$$

then

$$uv - ab \le \int_a^b U(z)\,dz + \int_a^b V(z)\,dz,$$

where $b = U(a)$.

If the lower limit of the integrals equals zero, this inequality becomes

$$uv \le \int_0^b U(z)\,dz + \int_0^b V(z)\,dz,$$

and it was in this form, with certain sharpenings, that it was subsequently applied.[85]

---

[85] Cf. Birnbaum and Orlicz [2, pp. 15–17]; Krasnosel'skii and Rutickii [1, pp. 23–24].

3.9 Strong and weak convergence

Young's note [4] was very short, and it is difficult to judge how clearly he conceived these classes of functions at the time. It is clear, however, that along with the idea of introducing the classes themselves in this note, he also generalized the concept of conjugate exponents, as can be seen, for example, from his own words:

> With this inequality [Young's inequality—F. M.] in our minds we may conceive of the summable functions grouped in pairs of classes in such a manner that *the product of two functions, one from each class of such a pair, is summable* [4, p. 225].

Young also found the conditions he was seeking for a given continuous function $F(x)$ to be the indefinite integral of an integrable function of a given class [4, pp. 227–228] and pointed out certain applications in the theory of trigonometric series.

Young developed his reasoning in more detail in his 1923 paper [7], which was published in 1926, but even this paper contains much that is unclear. The lack of clarity results mainly from the peculiarity of Young's works already mentioned. A more substantive analysis of this and other related works of Young would require a great deal of time and space, and would provide material for a separate study. We shall limit ourselves to one theorem of [7].

Let
$$Q(u) = \int_0^u U(z)\,dz,$$
where $U(u)$ *is a positive monotonically increasing function of u defined on $(0, +\infty)$ such that* $\lim_{u\to\infty} U(u) = +\infty$; *let* $\{f_n(x)\}$ *be a sequence of measurable functions.*

If
$$\int_a^b Q\{|f_n(x)|\}\,dx$$
exists for all values of n and is a bounded function of n, then in the sequence of integrals
$$\int_a^x f_n(x)\,dx$$
or in any subsequence of it, there exists a sequence converging to an integral, say
$$\int_a^x f(x)\,dx,$$
where $f(x)$ is of the same class of supersummable functions[86] *as the functions*

---

[86] Young's supersummable class is the set of $f(x)$ such that $\int_a^b Q\{|f(x)|\}\,dx < +\infty$.

that generate it, i.e., the integral
$$\int_a^b Q\{|f(x)|\}\,dx$$
exists.

Young's language did not become the generally accepted language of the theory of functions. The language of Riesz in [7], in contrast, became widely used, especially after his results relating to weak and strong convergence were expounded in the 1926 book of Hobson [3, pp. 249–257]. Still, Young's conceptual arguments were of undoubted interest, and Burkill [2] discussed them in 1928. For the sake of brevity we shall describe the contents of this work of his, using the terminology introduced in 1931 by Birnbaum and Orlicz [2].

A function $N(u)$ is called an *N-function* if it possesses the following properties:

1) it is defined and continuous for all u in the interval $(-\infty, +\infty)$;

2) $N(0) = 0$, $N(u) > 0$ for $u > 0$, and $N(-u) = N(u)$;

3) there exist numbers $\alpha > 0$, $\beta > 0$ such that for $u > \alpha$ we always have $N(u) > \beta$.

A measurable function $f(x)$ defined on $[0,1]$ is called *integrable with the N-function* $M(u)$ if
$$\int_0^1 M[f(x)]\,dx$$
is finite.

The N-function $N(v)$ is called the *conjugate* of the N-function $M(u)$ with respect to integration if

1) the integral $\int_0^1 f(x)g(x)\,dx$ is finite when $f(x)$ is integrable with $M(u)$ and $g(x)$ is integrable with $N(v)$;

2) a given function $f(x)$ is integrable with $M(u)$ if the integral
$$\int_0^1 f(x)g(x)\,dx$$
is finite for every function $g(x)$ that is integrable with $N(v)$.

A sequence of integrable functions $[f_n(x)]$ is said to be *strongly convergent with the N-function* $M(u)$[87] if
$$\lim_{\substack{p\to\infty\\q\to\infty}} \int_0^1 M[f_p(x) - f_q(x)]\,dx = 0. \tag{4}$$

---

[87] Birnbaum and Orlicz spoke of *convergence in mean* [2, p. 47].

A sequence of functions $\{f_n(x)\}$ converging with $M(u)$ is said to be *weakly convergent* with $M(u)$ to the function $f(x)$ if the inequalities

$$\int_0^1 M[f_n(x)]\,dx \leq K \quad (n = 1, 2, \ldots)$$

hold, where K is independent of n, and for every $t \in [0,1]$

$$\lim_{n \to \infty} \int_0^t f_n(x)\,dx = \int_0^t f(x)\,dx.$$

All the definitions just given, taken from the paper of Birnbaum and Orlicz, were present in the papers of Burkill, though not stated in such a didactically transparent form, to be sure. But, following Young, Burkill imposed on N-functions the restriction that they be indefinite integrals of positive functions defined on $(0, +\infty)$. The basic theorems of Riesz on strong convergence are generalized in his works. We shall give a detailed discussion of just one of them, namely the theorem on the approximation by step functions of functions that are integrable with an N-function $M(u)$. Riesz [7, p. 451–452] had proved this theorem for an arbitrary function in L^p. In generalizing it to this case Burkill neglected to impose an additional restriction on the growth of the function $M(u)$ consisting of the condition $M(2u) = O\{M(u)\}$ as $u \to \infty$. As a result of this his statement turned out to be wrong, as Birnbaum and Orlicz [1] pointed out in 1930 by constructing a suitable counterexample while sharpening the theorem and giving a new proof of it, moreover for continuous functions as well as step functions.[88]

Kaczmarz and Nikliborc [1] arrived at similar ideas at the same time as Burkill. However, while Burkill had started from Young's observations, the latter were responding to Noaillion's investigations,[89] which, as it happened, were never published and about which they had heard from Banach. Of their results we shall discuss in detail only the connection between strong convergence with an N-function and convergence in mean and in measure.

Kaczmarz and Nikliborc proved the following two theorems.

1. *Let $\{f_n(x)\}$ be a sequence of measurable almost everywhere finite functions converging almost everywhere on $[0,1]$. Further suppose there exists an N-function $g(x)$ satisfying the condition $g(x + y) \leq M[g(x) + g(y)]$ for*

---

[88] For the later investigations connected with this theorem cf. Ul'yanov [2].

[89] P. Noaillion. We have no information on this mathematician except his o. paper devoted to the Stieltjes integral.

every pair of numbers (x,y) (where M is a positive number) and such that the integrals

$$\int_0^1 g(f_n)\,dx, \quad \int_0^1 g(-f_n)\,dx$$

exist. Suppose finally that there exists a constant A such that

$$\int_0^1 g(f_n)\,dx < A.$$

Then the original sequence converges strongly with any N-function $g_1(x)$, satisfying the condition $g_1(x+y) \leq M(g_1(x) + g_1(y))$ for which

$$\lim_{|x|\to\infty} \frac{g_1(x)}{g(x)} = 0 \quad [1, \text{pp. } 161\text{--}163].$$

2. *Let $\{f_n(x)\}$ be a sequence of measurable almost everywhere finite functions converging in measure on $[0,1]$ to a function $f(x)$. Then there exists an unbounded function $g_1(x)$ of the same form as in the conclusion of Theorem 1 with which this sequence converges strongly* [1, pp. 163–166].

The authors stated and proved the second of these theorems for an unbounded N-function $g_1(x)$; in a footnote on p. 163 they noted that Young had proved the existence of a bounded function of this type in 1913.

In 1931 Birnbaum and Orlicz [2] undertook a systematization and didactic reworking of many of the studies on strong and weak convergence, supplementing it with their own numerous results, one of which we have already mentioned. There is hardly any need to restate these results here, and we shall not do this, but rather bring to an end our schematic survey of the history of the introduction and development of the concepts of strong and weak convergence, which is far from complete, even up to 1931. These kinds of convergence, when suitably generalized, soon grew into weak and strong convergence in more general functional spaces, in relation to which we refer the reader to the books of Banach [2, pp. 106–124, 207–209] and Kantorovich and Akilov [1, pp. 270–284].

In addition to this, we also decline to consider other forms of convergence of sequences of functions, such as absolute, monotone, and unconditional convergence, although they play an important role in the theory of functions, comparable to the role of the kinds of convergence we have considered. Every significant problem of the theory of functions usually has a large and complicated history; to trace that history is a difficult and thankless task.

3.10 The Baire classification

In the discussion of the comparison of convergence everywhere and convergence almost everywhere in Sec. 5 of this essay what was said might lead one to believe that convergence everywhere is of no importance. This, however, is true only to a certain degree and only in the metric theory of functions. If we turn to descriptive function theory, we encounter a situation that is nearly the exact opposite: in descriptive function theory, as a rule, convergence almost everywhere is unimportant, and the main tool of investigation is provided by convergence everywhere and even less general forms of convergence. In order to illustrate this assertion, let us return, for example, to the theorem of Fréchet mentioned above [4, pp. 16–17]. This theorem says that every measurable almost everywhere finite function—until recently and still to a great degree the principal object of study in the theory of functions—can be represented as the limit of an almost everywhere convergent sequence of continuous functions. In other words, starting from the class of continuous functions and using convergence almost everywhere we can obtain any measurable function of the class of measurable almost everywhere finite functions. This by itself closes off the road to the study of intermediate classes of functions; and, for example, one cannot distinguish the functions of the first class in the Baire classification, to which the most useful functions of analysis belong—all the ordinary derivatives of continuous functions, the integrals of partial differential equations, monotone functions and functions of bounded variation, etc. Such a picture arises not only in the question of the analytic representation of functions, but also in many other questions, where convergence almost everywhere is too general and does not allow the possibility of obtaining an answer to an interesting question.

Convergence everywhere, in contrast, makes it possible to obtain immediately from the continuous functions both the functions of the first Baire class and—by iteration—the functions of higher classes, to distinguish objects of study that deserve the most intense scrutiny. We shall take up these questions briefly in the present section.

When sequences of continuous functions were studied in the nineteenth century, mathematicians encountered two sets of facts: some sequences led to functions that were also continuous, while discontinuous functions were obtained using other sequences. At first the efforts of mathematicians were concentrated mainly on characterizing in some manner the everywhere convergent sequences of continuous functions that yielded continuous functions as a limit. These efforts, as was shown in Secs. 3 and 5, were crowned by Dini's Theorem giving conditions for the continuity of the limit at a point, on the one hand, and by Arzelà's Theorem on quasiuniform convergence on the other.

The problem of characterizing the class of discontinuous functions that are limits of everywhere convergent sequences of continuous functions came onto the agenda. It was this problem that Baire undertook to study starting in 1897.[90]

Baire [1] began with the problem of the connection between the continuity of a function of two variables in each argument separately and its continuity jointly in the two variables. Introducing the concept of a semicontinuous function[91] and the oscillation of a function at a point [1, p. 694], he solved the problem just posed using these concepts, in the sense that every function of two variables that is continuous in each argument separately is pointwise-discontinuous with respect to the two variables jointly, i.e., on any curve in the domain of definition of the function there exist points at which it is jointly continuous in the two variables [1, p. 693].[92] In particular, he obtains from this the result that a function satisfying the hypotheses of the preceding theorem yields a function of one variable on the line $y = x$ that is pointwise-discontinuous on that line; and here the problem is posed of characterizing the class of pointwise-discontinuous functions of one variable. In [1] Baire succeeded in obtaining only certain particular results. Evidently the principal reason for his lack of success was an insufficiently general approach to the definition of semicontinuity and to the concepts of oscillation and pointwise discontinuity of a function, which were studied in [1] only for intervals.

But by the following note [2], having defined these concepts in relation to an arbitrary perfect set, he obtained the fundamental result, now known as Baire's Theorem on functions of first class:

If a series whose terms are continuous functions of x converges for each value of x, it represents a function that is pointwise-discontinuous with respect to every perfect set. Conversely, if a function $f(x)$ is pointwise-discontinuous with respect to every perfect set, then there exists a sequence of continuous functions $f_1(x)$, $f_2(x)$,..., $f_n(x)$,... that converges to $f(x)$ for every value x_0 of the variable x [2, p. 886].

In this way the first class of functions in the Baire classification was marked out, and a structural characterization of the functions of this class was given. Because of its significance, noted above, this theorem immediately

---

[90] For the life and works of Baire cf. Dugac [2].

[91] Cf. Natanson [1, Vol. II, p. 149].

[92] As noted earlier, the concept of a pointwise-discontinuous function goes back to Hankel [2] and was the principal object of study in investigations in the theory of functions until the time of Lebesgue.

caught the attention of mathematicians. Baire himself, in [2], merely indicated an outline of the proof of the theorem just quoted. In the following year he devoted almost the entire second chapter of his dissertation [5, pp. 19–63] to it, and then returned to its proof at least three more times. Lebesgue and Young each gave several proofs of this theorem, as did many mathematicians after them. The history of this theorem, its modification and generalization, could form the subject of a large and interesting separate article. In the present section, we have set as a goal only a brief description of the Baire classification, and therefore we now turn to that.

Baire published his classification [3] two and a half months after his characterization of the functions of the first class. He assigned the continuous functions defined on $[a, b]$ to the zeroth class. If $f(x)$ is representable as the limit of an everywhere convergent sequence of continuous functions, i.e.,

$$f(x) = \lim_{n \to \infty} f_n(x) \qquad (1)$$

at each point $x \in [a, b]$ for continuous functions $f_n(x)$, and is not continuous, then this $f(x)$ is a function of first class. The functions belonging to neither the zeroth nor the first class but representable in the form (1), where $f_n(x)$ belongs to the zeroth or first class, form the second class, etc., for all classes with finite index. But Baire continued this classification even further. About this he temporarily writes very laconically:

> It is possible, however, to go further, using the concept of a transfinite number. If there is a sequence of functions, each of which belongs to one of the classes $0, 1, 2, \ldots, n, \ldots$, and there exists a limiting function that does not belong to any of these classes, we shall say that this limiting function belongs to class ω. We further posit the possibility that there exist functions belonging to class α, where α is any transfinite number of second class [3, p. 1662].

A detailed exposition of this classification was given by Baire in his 1899 dissertation [5, Ch. III]. This dissertation became one of the most important milestones in the development of the theory of functions. Although the basic ideas contained in it had been for the most part expounded in his preparatory works already mentioned, they were presented most fully here, and it is with this work of Baire's that many mathematicians subsequently began. It suffices to note that such concepts as the concept of a semicontinuous function [5, p. 10] and the concept of sets of first and second category and the essential difference between them [5, pp. 65–66], etc. were introduced here. A detailed proof of the theorem on functions of first class was given [5, Ch. II]; the set-theoretic method is consciously made the basis of the entire exposition, and along with it the critical approach to the principles of set theory is extended, an approach which in a few years was to grow into a serious theoretical battle in the area of foundations of mathematics. We shall give a somewhat more detailed discussion of Baire's attitude toward transfinite numbers, since they bear directly on his classification.

Cantor [5] had introduced transfinite numbers in 1880, and again [7] in 1883, and had given yet another foundation for the theory of them [8] in 1897.[93] The introduction of these numbers had evoked many objections. But Baire needed them, and he adopted the following position:

> In this regard I remark once and for all that we shall not occupy ourselves with the difficulties to which the abstract concept of a transfinite number by itself leads, although this expression will be used in our work. In the present case, for example [he is discussing the introduction of the concept of a derived set of order α, where α is a transfinite number of second class—F. M.], the sets P^α, where α is a definite number of the second number class, is a completely definite thing, independently of any abstract considerations regarding the symbols of M. Cantor; consequently what is involved is merely the use of ordinary language [for naming things like the set P^α—F. M.] [5, p. 36].

Baire understood that his classification would be incomplete if limited only to natural numbers; moreover he had been unable to prove even his basic theorem on functions of first class without using transfinite numbers.

The introduction of the Baire classification immediately posed a host of diverse questions for mathematicians.

A collection of functions that we shall denote by the letter B is defined by this classification. The first natural question is, could one obtain a function not belonging to the class B by applying the same limiting passage to functions of the class B?

Baire gives an immediate answer to this question: if a sequence $\{f_n(x)\}$, with $f_n(x) \in B$, converges everywhere to some function $f(x)$, then $f(x) \in B$ [3, pp. 1622–1623; 5, p. 70]; this answer is obtained as a simple corollary of Cantor's theorem that every countable sequence of numbers of first or second kind has an upper bound that is a number of class two at the highest. Consequently the family of functions B is closed with respect to this form of convergence.

But then a second question is natural: Doesn't the collection B then encompass in general all conceivable single-valued functions defined on $[a, b]$?

---

[93] For more details on these papers of Cantor cf. Medvedev [2, pp. 117–125, 171–178].

3.10 The Baire classification

For the end of the nineteenth century this was not an idle question. Baire discovered that the set B has cardinality of the continuum [3, p. 1623]. In this connection it had long since been known that the set of all real-valued functions defined on $[a, b]$ has cardinality 2^c. Hence the set B is only a small subset of the set of all functions.

The following problem turned out to be more difficult. Baire, as mentioned, had succeeded in obtaining a structural characterization of the functions of first class. It was a legitimate goal to find a similar characterization of the functions of higher classes. Baire first discovered that the functions of second class are pointwise-discontinuous on every perfect set, neglecting a set of first category with respect to the perfect set [3, p. 1623];[94] he proved this property of functions of second class in his dissertation [5, pp. 81–87], and soon established that it is a property not only of functions of second class, but in general of all functions belonging to his classification [4]. The question posed in [4], whether this necessary property of Baire functions was also sufficient, was a question that Baire did not succeed in solving. A negative answer was given in 1914 by Luzin [3], who used the continuum hypothesis to construct functions possessing this necessary property but not belonging to the collection B. Luzin [5] later confirmed this result of his without applying the continuum hypothesis, but using the axiom of choice, and in a joint paper with Sierpiński [1] in 1928 proposed an effective example of such a function.[95]

The fact that functions of the family B constitute only an insignificant portion of the set of functions in general might lead one to suspect that the formal definition of the various Baire classes is sterile in the sense that all the classes are empty from some index on. The existence of functions of the zeroth and first classes was known for certain, and the well-known function of Dirichlet [1, p. 23] was an example of a function of second class. What the situation was from that point on was still unclear in 1904, when Lebesgue [9] established the existence of functions in every Baire class.

This note of Lebesgue's [9] was only a brief preliminary report on his important memoir "Sur les fonctions représentables analytiquement" [10], in which the Baire classification received a profound further development. In 1930 Luzin wrote of this memoir, "...this memoir of Lebesgue's, which appeared in 1905 and is extremely rich in ideas, methods, and general results, is still far from being completely studied" [6, p. 298]. To a certain degree this can still be said today, although a very large number of mathematicians have

---

[94] It was in this paper of Baire's that sets of first and second category were introduced. For brevity we give Baire's results in Lebesgue's formulation [10, pp. 184–188].

[95] The term *effectiveness* has many meanings, and we shall not make it more precise at this point. For a modern interpretation of Luzin's results cf. Uspenskii [1, p. 103].

studied it. By historians of mathematics it has not been studied at all, except for particular citations of certain factual results contained in it. We also do not propose to make such a study, but will limit ourself to particular remarks.

Shortly before the publication of Lebesgue's memoir there appeared the memoir of Zermelo, "Beweis, daß jede Menge wohlgeordnet werden kann" [1], which contained a statement of the famous axiom of Zermelo (the principle of arbitrary choice) and a proof based on it that every set can be well-ordered. Mathematicians had long used similar reasoning implicitly, but when Zermelo conducted such reasoning explicitly, he caused a storm in mathematical circles. An acrimonious discussion immediately began on the principles of mathematics in general, in which the participants were Hadamard, F. Bernstein, Borel, Baire, Lebesgue, Russell, Zermelo, and others.[96] This discussion left a deep impression also on the memoir of Lebesgue we are discussing.

This was shown first of all by the fact that Lebesgue made virtuoso attempts to avoid any reasoning involving the axiom of choice. To do this he developed a method of constructing point sets based on the use of the set of transfinite numbers of second kind; moreover the transfinite sequences of real numbers occurring in the set under consideration are defined not simultaneously as an actual set, but successively, so that the definition of each number of the sequence depends essentially on the definition of the preceding numbers. In this way each of the sets considered by Lebesgue is in a certain sense constructive.[97]

Lebesgue, following Borel and Baire, refused to consider the transfinite numbers themselves in the form in which they had been introduced by Cantor in [8]. For Lebesgue these numbers were mere symbols for denoting certain classes of sets or functions that can be constructed from classes already constructed [10, p. 148]. As it happens Cantor had originally introduced them this way [5, pp. 147–148], in the form of symbols to characterize the derived sets of various orders of a given set; Borel [2, pp. 140–141] had approached them in the same way in 1898, introducing them in order to characterize the same sets and orders of growth of functions; Baire expressed the same point of view on them, as already said. With Lebesgue this approach assumed a more finished form, and was developed into a convenient algorithmic method

---

[96] Part of this discussion is described by Luzin [7, pp. 509–512], but many others have also described it. We mention only our own book [5, pp. 105–113], in which, by the way, the contents of the present *Essays* are also supplemented.

[97] Cf. Luzin [6, pp. 298–299]. The term *constructive* is used in very diverse ways by mathematicians. Here a constructive definition of a set is understood as a definition not involving the axiom of Zermelo or the totality of all transfinite numbers of second kind, considered as an actually realized set.

3.10 The Baire classification   163

of studying classes of sets and functions using mostly transfinite induction.

This approach to transfinite numbers is of historical importance. The theory of sets, as it was formed in the nineteenth century and at the turn of the century was strongly discredited both by the paradoxes uncovered within it and by the unsatisfactory state of its foundation from the point of view of the increasing demand for rigor. There existed a real danger that mathematicians would turn away from the theory that had been so compromised in the discussions at the beginning of the century. The great merit of Borel, Baire, and Lebesgue was that they adopted abstract set theory as a tool, adapting it to satisfy the specific needs of the theory of functions. And Fréchet was perhaps not far from the truth when he named the works of the French scholars on the transformation of the concept of a transfinite number among the fundamental stages in the development of the subject of sets, comparable in significance to the creation of set theory itself by Cantor [7, p. 15].[98]

Another important peculiarity of Lebesgue's memoir [10] was a closer connection between investigations in the theory of point sets and the theory of functions than is found in Borel and Baire. The connection between them existed from the very moment of their first appearance, and was recognized in completely general form by Baire as early as 1899, when he wrote that all problems involving functions reduce to certain problems of set theory [5, p. 121]. This connection was widely used by Borel, Baire, Lebesgue, and others in numerous investigations at the end of the nineteenth century and the beginning of the twentieth. Very many of Baire's results on the study of functions of first class and his classification of functions were based on this connection. But by 1905 a curious situation had arisen.

As early as 1898 Borel [2, pp. 46–49] had introduced the class of sets now known as Borel sets, or simply B-sets. On the other hand, in that same year Baire had begun to study the class of functions belonging to his classification and now called Baire functions or simply B-functions. The two scholars knew each other's works very well, but in general it could be said that for them the two mathematical objects—B-sets and B-functions—had no connection with each other; in particular there was no classification of B-sets similar to the classification of B-functions. Simultaneously with the writing of Lebesgue's memoir [10], the books of Borel [7] and Baire [7] were being prepared for printing, and all three works appeared in 1905. But neither Borel, studying B-functions, nor Baire, referring to the works of Borel, noticed the kinship between B-sets and B-functions. The establishment of this kinship is one of the most important achievements of Lebesgue.

In connection with the introduction of the new concept of an integral in

---

[98] This transformation was subsequently continued; cf., for example, Luzin [6, pp. 29–30].

1901 Lebesgue [2] had introduced an extremely fruitful method of studying functions, which consisted of studying the sets $E[f(x) > a]$, i.e., the set of $x \in E$ at which the values of a function $f(x)$ defined on the set E exceed a certain real number a. This method rendered essential assistance to him in the introduction and study of the new concepts of measure and integral, as well as the concept of a measurable function, and in the study of a large number of problems of the theory of functions. Lebesgue was not the first to apply it, and in a certain sense its lineage can be connected with the integrability condition stated by Riemann. Harnack, as we have said, also used such sets, and Baire had applied them. But Lebesgue went much further than his predecessors, developing a general method that is widely applied even today. The generality in which Lebesgue conceived this method can be seen from just the following words of his: "For me defining a function means defining a correspondence $[f(x), x]$, which in turn means defining the sets $E[a < f(x) < b]$ " [18, p. 236].[99] It rendered invaluable assistance to Lebesgue in [10] also.

It would take too long to describe the details of Lebesgue's approach to the connection between B-functions and B-sets, although these details are of some interest. We shall state only some of his basic results in this direction.

In his dissertation [3, pp. 257–258], and then in his book [8, pp. 111–112] Lebesgue established that every B-function $f(x)$ possesses the property that for it the sets $E[f(x) > a]$ are B-sets. At the time, however, this was for him, as for others, merely stating a mathematical fact. But in the memoir "Sur les fonctions représentables analytiquement" this fact grew into a classification of the B-sets: a B-set $E[f(x) > a]$ corresponding to B-functions of class α is precisely a set of a completely definite class of Borel sets, to which one naturally ascribes the same index as the function. In this way the first classification of the B-sets appears, to whose study Lebesgue devoted the fourth section of his memoir [10, pp. 156–166]. Subsequently this classification was studied by many mathematicians, and modifications of it of greater and lesser importance were given, of which one may, for example, mention the classifications of Hausdorff and Vallée-Poussin and the modification and deepening of the latter by Luzin [6, pp. 53–134].[100]

The introduction of the classification of B-sets enabled Lebesgue to prove the following theorem: *A necessary and sufficient condition for a function f defined everywhere to be of class α is that for any numbers a and b the set $E(a \leq f \leq b)$ be of class α or lower and that it be actually of class α for some a and b* [10, p. 167].

---

[99] The fact that the quotation involves the sets $E[a < f(x) < b]$ and not the sets $E[f(x) > a]$ is unimportant; Lebesgue had shown this already in his dissertation [3, pp. 250–251].

[100] For these cf., for example, Shchegol'kov [1, pp. 20–43].

3.10 The Baire classification

The memoir of Lebesgue under consideration is interesting not only for establishing the connection between B-sets and B-functions. As mentioned, Baire had given a structural characterization of the functions of first class, but had not succeeded in doing this for functions of higher classes. Lebesgue managed to find structural properties of functions of an arbitrary class. In order to state them we shall also need Lebesgue's definitions.

He defined a point set F to be a set of class α if it can be regarded as a set $E[a \leq f \leq b]$ corresponding to a function $f(x)$ of class α but not for a function belonging to a class of index lower than α [10, p. 156].

According to Lebesgue a *set of rank α* is defined as any set that can be regarded as the intersection of a finite or countable number of sets F belonging to classes with indices less than α, while no such representation exists when the number α is replaced by a smaller number [10, p. 161].

The function $f(x)$ is called a *B-function within ε* if there exists a B-function $\varphi(x)$ such that $|f(x) - \varphi(x)| < \varepsilon$ for all x. If $\varphi(x)$ can be chosen so that it belongs to class α but not so that it belongs to a class with index less than α, then $f(x)$ is called a function of class α within ε. In particular, if $\alpha = 0$, i.e., $\varphi(x)$ is continuous, then $f(x)$ is called continuous within ε; when $\varphi(x) = \text{const}$, the function $f(x)$ is called *constant within ε* [10, p. 170].

A function $f(x)$ is called *continuous at the point $P \in E$ neglecting sets of a family in E* if there exists a certain set e of this family such that $f(x)$ is continuous at P on the set $E - eE$.

Finally, $f(x)$ is said to be *pointwise-discontinuous on E neglecting the sets of a certain family of sets in E* if E is the derived set of the set of points in E at which $f(x)$ is continuous neglecting the sets of this family [10, p. 185].

We can now state the definition and theorem of Lebesgue that Luzin [6, p. 302] called the culmination of this paper of Lebesgue.

A function $f(x)$ is said to be continuous (α) at the point P of a perfect set E if to every $\varepsilon > 0$ one can assign an interval containing the point P on which E can be regarded as the union of a countable set of sets of rank at most α, on each of which $f(x)$ is constant within ε, and all the latter sets except those containing the point P are nowhere dense on E. If E is the derived set of the set of points at which $f(x)$ is continuous (α) on E, then $f(x)$ is called *pointwise-discontinuous (α)* on E [10, p. 191].

> In order for the function f to belong to class α at most, it is necessary and sufficient that it be pointwise continuous (α) on every perfect set [10, p. 191].

This last theorem is a precise analogue of Baire's theorem on functions of first class. Besides this theorem Lebesgue proved several other theorems giving a structural characterization of functions of class α. We give another formulation of it: *A necessary and sufficient condition for a function f to be of class $\alpha > 0$ is that for any ε the domain of definition of f can be regarded as the union of a countable set of sets of rank at most α, on each of which f belongs to a class less than α within ε (or on each of which f has oscillation at most ε)* [10, p. 173].

Much more could be said about Lebesgue's memoir [10],[101] but we shall add only the following to what has already been said. Here he established the existence of functions of any Baire class (pp. 205–212), and also constructed an example of a function not belonging to this classification (p. 216). Lebesgue's proof of the existence of such functions was noneffective, in the sense that the actual construction of a function of the class in question was not carried out. The difficulty of the construction can be seen from the fact that, while an arithmetic example of a function of the third Baire class was constructed in 1906,[102] a quarter of a century passed before a similar example of a function of the fourth class was constructed by L. V. Keldysh.[103] We note also that an error of Lebesgue, consisting of the assumption that the projection of a *B*-set is also a *B*-set [10, pp. 191–192], led him to certain incorrect or incorrectly stated propositions, which he corrected in a 1918 paper [18, pp. 241–243], after Suslin discovered his error.

As it turned out there were innumerable problems that arose with the appearance of the Baire classification of functions and the classification of sets proposed by Lebesgue. If we were to attempt to trace the subsequent statement and solution of these problems, we would have to raise too large a circle of questions connected with descriptive set (or function) theory, which in turn is interwoven with functional analysis and topology, mathematical logic and the philosophical questions of mathematics as a whole. For that reason we shall call a halt and in conclusion make only the following observations.

The basis of the Baire classification of functions or the Lebesgue classification of *B*-sets is passage to the limit of a sequence at every point. Naturally the various kinds of uniform limiting passages, including Arzelà's quasiuniform convergence were of no use in investigations of this type, simply because none of them led beyond the boundaries of the class of continuous functions, although the last of these types of convergence, when suitably generalized, was used to establish criteria under which a convergent sequence of functions of a given class has a limit belonging to a class that is no higher than the given class. But two other kinds of nonuniform convergence of sequences of functions that are special cases of convergence everywhere, but without the narrowness

---

[101] A portion of Lebesgue's results were studied by Luzin in [6], cf., in particular, pp. 298–320.
[102] For this example, cf. Luzin [6, pp. 92–93].
[103] For this example cf. Luzin [6, pp. 97–104].

3.10 The Baire classification

of uniform convergence, with which it is not possible to pass outside the domain of continuous functions, had long been applied in mathematics, namely monotone convergence and absolute convergence.

Monotone increasing and monotone decreasing sequences of functions had been used in numerous investigations by W. H. Young. He had in fact constructed a classification of functions based on limiting passages of this type [3, pp. 15–24] and used it in this same paper to construct another theory of integration equivalent to that of Lebesgue. However, the idea of such a classification was in a sense contained in a paper of Baire [6], in which the theorem was proved that $f(x)$ is the limit of a monotone increasing (resp. decreasing) sequence of continuous functions if it is upper (resp. lower) semicontinuous[104] and also the theorem that the limiting function of a monotone decreasing (resp. increasing) sequence of upper (resp. lower) semicontinuous functions is upper (resp. lower) semicontinuous.[105] Young seems to have arrived at similar results independently of Baire and somewhat later. But he went further, using the fortunate idea of combining decreasing and monotone increasing limiting passages. By this route Young obtained a new classification of functions (and later sets) similar to the Baire classification in many ways, but at the same time different. This classification, both intrinsically and in relation to the Baire classification was also the source of many and various investigations that continue down to the present time.

The idea of using absolutely convergent series of functions to obtain functions of the family B was stated by Sierpiński [1] in 1921. Having established that a necessary and sufficient condition for a function of a real variable $f(x)$ to be representable by an absolutely convergent series of continuous functions is that $f(x)$ be the difference of two semicontinuous functions, and having established that there exist functions of the first Baire class that are not expandable in such a series, he proposed the idea of a new classification. The study of this classification was immediately taken up by S. Mazurkiewicz [1] and especially Kempisty [1]. Sierpiński's classification seems to have attracted less attention than that of Young, and especially that of Baire.

Each of these classifications, both individually and in relation to the others, has been studied by a large number of scholars. The problems that arise in one of them have been transferred to the others, and the solutions have sometimes appeared to be different for different classifications. To describe them would require time and space and so we limit ourselves to what has already been said and pass to another question of the history of the theory of functions.

---

[104] Cf. Natanson [1, Vol. II, pp. 153–156]; Baire had in mind bounded functions.

[105] Cf. Natanson [1, Vol. II, pp. 151–152].

Chapter 4
The derivative and the integral in their historical connection

4.1 Some general observations

The two preceding essays are constructed in general, so to speak, according to the principle of microscopic examination of particular details in the development of certain ideas of the theory of functions, at least in the earlier stages. Such a method has its advantages, enabling the author to concentrate on the points of interest and trace the course of one stream of thought or another whose confluence forms a broader stream of ideas. With this method it becomes possible to approach nearer to the causal connections and trace the role not only of the leading scholars but also the toilers in the vineyards of science whose activity is not only useful but also necessary. For a correct understanding of the course of historical development this is essential, for "the history of science does not reduce to the exposition of only the ideas of the first-rate minds... the progress of science depends no less on the way in which these epoch-making discoveries are prepared for by the day-to-day labor of many hundreds of scholars of lesser rank, the way in which particular problems gradually lead to the posing of general problems, the creation of general methods of investigation, and the formation of general scientific concepts" (Gnedenko, [1, p. 11]).

At the same time this method has many defects. First of all, the field of vision remains very restricted and the general picture of the development is lost, since a situation arises that is very well described by the cliche "not seeing the forest for the trees." In addition, tracing the details requires the writer to bring in such a large volume of literature that this approach turns out to be unrealizable in practice without restricting oneself to comparatively minor themes—in this sense it turned out to be unrealized in the preceding essays because of many circumstances: the practically unsurveyable scientific literature, which incidentally is frequently very difficult of access; mathematical and linguistic difficulties; the desire to reduce the references to a reasonable minimum, and the like. Finally, it is far from being the case that every reader is interested in piling up details, even necessary details.

Such an approach is even more unrealizable in the case of the theme "differentiation and integration." It suffices to say that a bibliography of it would occupy the entire volume of the present essay, even if held to a reasonable minimum. It therefore seems reasonable to choose an interesting and important idea for our *Essays* and trace its development in broad outline over the largest period possible.

We have chosen the idea of the connection between the two basic infinitesimal operations as this idea. It is interesting and important if only because classical analysis basically owes its origin to this connection. In addition the general vicissitudes of this connection are extremely curious—from a separate development of the concepts of derivative and integral through their mutual connection when the operation of differentiation came to the fore, then the subsequent transition to a generally separate development of these concepts, but with a constant tendency to to revive the lost connection, and finally the manifestation of a tendency to reverse the relationship between derivative and integral, which led to the appearance of the Newton-Leibniz-Euler analysis, i.e., to place integration in the foreground and define differentiation in terms of it. Such metamorphoses of this connection were evidently somehow caused by the general course of development of mathematical science, but to trace the latter circumstance is too complicated a problem, and we decline to make the attempt to solve it. In what follows we shall give only particular hints as to the cause of the transformation of the relationships between the derivative and the integral through certain trends in the development of science.

It is also of some interest that the purely mathematical development of the idea of the original introduction of a definite integral and the definition of the concept of a derivative in terms of it was preceded by the methodological idea of an earlier introduction of the former in the teaching of mathematical analysis, the derivation of the properties of the definite integral not from properties of the derivative, as had usually been done during the classical period, but from a direct definition of it. In this connection a rather lively polemic took place, a discussion of which would draw us away from our main topic, and therefore we shall have to restrict ourselves again only to remarks made in passing.

The tendency to study first the concept of a definite integral and then define the concept of a derivative in terms of it appeared quite distinctly in both its purely scientific and its methodological aspects. However it by no means became predominant, and it is not clear whether it will become so. In modern investigations one can detect almost all the historically existing types of relations between differentiation and integration, but a study of them would take us too far beyond the scope of this book, as outlined in the introduction. The reason is that the study of the concepts of derivative and integral and the operations of differentiation and integration, is nowadays carried out mainly on the level of functional analysis, and the study of the various types of relations and a comparison of them would require us to turn to the history of this enormous mathematical discipline, which we as a rule prefer not to do.

And now a final remark. Since we intend to trace the idea we have chosen over a long period of time, and at the same time we decline to give a detailed discussion of the vagaries of its development in the whole variety of its transformations, we shall very frequently rely, partly for brevity and partly because of ignorance of the relevant literature, on historico-mathematical studies, but of course only if the question under investigation has been elucidated in a form that we find satisfactory in the historico-scientific studies known to us.

4.2 Integral and differential methods up to the first half of the seventeenth century

There is no unanimity of opinion among historians of mathematics on the question of the time at which integral and differential methods arose. Since we are not interested here in the question of the origin of these methods but only in their mutual relation, we shall not only note the divergence of opinion among historians and attempt to explain this divergence, but also choose a point of view and give the grounds for such a choice. We note only that we are assuming the possibility of assigning the appearance of integral methods to an era preceding the works of Archimedes, and our only argument in support of such a possibility is the following.

The creation of the method of exhaustion is usually ascribed to Eudoxus of Cnidos (*History of Mathematics...* [1, pp. 101–105]), who lived more than a century before Archimedes. Now the method of exhaustion was created and applied to solve problems that are now usually solved using integral calculus. And if we accept, along with the majority of historians of mathematics, that integration techniques were invented only in the works of Archimedes, we must admit a rather unlikely situation: to accept the possibility that a method of reasoning was created, a method of grounding the solution of a certain class of problems, before these problems or similar ones had begun to be solved at all. The usual situation in the history of mathematics was quite the reverse.

As for differential methods, the picture here is simpler.[1] At present we know of their existence only in the works of Archimedes (*History of Mathematics...* [1, pp. 124–128]).

Let us now discuss the mutual relationship of these two types of method.

---

[1] Or perhaps more complicated. The reason is that the study of differential methods in antiquity has only just begun, and they have been studied only in the works of Archimedes (Bashmakova [1]). It may be that they will also be discovered in other ancient mathematicians.

The first thing that should be done in this regard is to state the fact that they relate to two different classes of problems. Integral methods were applied in finding areas, volumes, and centers of gravity, while differential methods were used to determine tangents and extrema. These problems themselves seem to have no connection with each other. However, in the methods of solving them there was at least one common feature—the application of infinitesimal considerations, either explicitly or in disguised form. In the problems of the first type the unknown quantity was divided into infinitely small pieces, and in finding a tangent an infinitesimal, the so-called characteristic triangle, was introduced. But this feature was certainly not to any extent recognized at the time as a common characteristic of the two methods.

The second thing that should evidently be mentioned in connection with integral and differential methods in antiquity is that the extent of the class of problems that were solved by the former was significantly larger than the class of problems for which the latter were applied. While the application of differential methods was for the time being noted only in cases of finding the tangent to the spiral of Archimedes and the reduction of the problem of finding the maximum of the expression $x^2(a-x)$ to the problem of constructing a tangent, integral methods were applied rather widely. Thus, if we consider only quadratures obtained (or rather proved) using the method of exhaustion, there are some twenty-odd problems of this type in the writings of Archimedes and his predecessors.

The third point that deserves attention is that the infinitesimal methods of antiquity were applied to problems of a static character—the calculation of areas and volumes of unchanging figures or finding their centers of gravity.

A nearly constant relation between the number of problems that were solved by the two methods persisted in general for quite some time afterwards. The Archimedean quadratures and cubatures were supplemented somewhat by Pappus (cf. Medvedev [4]), Thābit ibn-Qurra, al-Kuhi, and ibn-al-Haitham (*History of Mathematics...* [1, pp. 238–243], Yushkevich [3]). Differential methods also underwent a further development, and we should pause to discuss this in more detail.

The enlargement of the circle of problems requiring the application of differential methods for their solution occurred outside of mathematics proper and was connected with the kinematical representations of astronomy. As early as the second century B.C.E. the Babylonian astronomers, in studying the motion of the heavenly bodies, essentially arrived at the concept of a function defined in the form of a table and studied the extrema of the functional dependences they were considering, actually calculating the value of the derivative at a point of maximum or minimum (Hoppe [1, pp. 150–151]). And in the second century C.E. this method of defining and studying functional dependences in astronomical problems was successfully applied by Ptolemy in optics (Schramm [1, pp. 72–80]).

The trend just described underwent a further development, again primarily in the framework of astronomy and optics, in the works of many Islamic scholars, especially in the works of the two greatest scholars of the Middle Ages (ninth to eleventh centuries)—ibn-Qurra and al-Bīrūnī. They took a more general approach to the concept of functional dependence and the study of extrema of functions, as well as to representations of instantaneous velocity and acceleration in the movement of heavenly bodies (Schramm [1], *History of Mathematics*... [1, pp. 241–242], Rozhanskaya [1,2,3]).

The same questions were considered more broadly in a certain sense by the scholars of Medieval Europe in the thirteenth and fourteenth centuries, especially Bradwardine, Swineshead, and Oresme. They studied a more abstract motion than the motion of the heavenly bodies in astronomy and attempted to give it a mathematical interpretation. In this connection the concepts of velocity and acceleration, distance and time of motion and along with them the concepts of variable quantity and function appeared in more general form (*History of Mathematics*... [1, pp. 270–282]).

Thus if infinitesimal methods found their original application in the framework of static problems, the circle of problems requiring infinitesimal techniques for their solution gradually enlarged, and this enlargement took place in such a direction that problems involving the application of infinitesimal methods occupied an ever larger—though for a while only relatively larger—place in the whole circle of infinitesimal problems and were connected with the notions of kinematics.

Thus one can say that in ancient and medieval times the rudiments of integral and differential methods were developed in application to different classes of problems. These methods were not connected with each other in any way, unless one counts a certain unifying characteristic of them—their infinitesimal character—which was not recognized by an official mathematics that was based on the method of exhaustion.

The excessively general and consequently indefinite approaches of the Western European scholars, associated with the scholastic philosophy, contained a variety of profound qualitative ideas whose realization could be achieved only after the development of the suitable mathematical apparatus. It is interesting that the mathematicians of Western Europe during the fifteenth and sixteenth centuries seem to have ceased to study infinitesimal problems unless one counts a few works on the determination of centers of gravity connected with translations of the works of Archimedes. It is as if they sensed that speculative arguments were not sufficient for progress, but that a preparatory development of computational algorithmic methods of science was needed. This is what they undertook in the fifteenth and sixteenth centuries (*History of Mathematics...* [1, p. 5]). But from the very beginning of the seventeenth century, after arithmetic and algebraic symbolism had been created and fractional and negative exponents and negative and complex numbers had been introduced, after algebra had been constructed in the form of a certain symbolic computation using letters for the coefficients and unknowns and the symbolism of algebraic operations had been extended, decimal fractions had come into common use, plane and spherical trigonometry had progressed, and methods of computing tables had been perfected—only after this did mathematicians again take up the study of infinitesimal problems, and significantly more intensively than in any preceding period.

At first these investigations had almost the same character in this period as in antiquity and in the Middle Ages: there were two classes of problems unconnected with each other, and the problems of the first class were solved using integral methods while those of the second class were solved using differential methods. The total number of such problems grew rapidly, and again one can establish an increase in the relative weight of problems involving differential methods connected with the study of motion.

However with time certain features of infinitesimal investigations that either had not existed at all earlier or had been in the background began to manifest themselves. We shall briefly discuss only a few of these.

Up to the beginning of the seventeenth century the legitimate method of reasoning in applying differential and integral techniques was the method of exhaustion. The characteristic property of investigations in analysis in the seventeenth century became the conscious rejection of this method of reasoning and its replacement by arguments based on infinitesimal notions in their various forms. Such a rejection was explicitly proclaimed by Kepler and Cavalieri, and after them many others began to proceed similarly despite the strong opposition of the adherents of traditional ways of thinking.

To be sure arguments connected with the application of infinitesimals can be found in the writings of Archimedes; Pappus made wide use of them; it is likely that the arguments of the Babylonians were a kind of equivalent of them, as were those of the Islamic mathematicians, astronomers and opticists in describing the celestial motions and optical phenomena and in analyzing the functional dependences defined by tables. But these arguments had long lain outside the realm of official mathematics. The introduction of them into the domain of legitimate mathematical thought had the significant consequence that it made it possible to combine according to the method of solution two classes of problems that had hitherto been kept distinct. Indeed the classical argument used in the method of exhaustion was based on the application of inscribed and circumscribed figures. For integral techniques such an approach was to some degree natural, since it was possible to inscribe (or circumscribe) approximating figures with known area or volume for the comparatively simple objects that were being studied at that time. But in solving problems of differential character (finding tangents and extrema), especially in studying motion, no such natural inscribed or circumscribed figure existed. For that reason differential and integral methods were to some extent opposed to each other. The transition to infinitesimal reasoning removed one of the essential differences between them and made it possible to combine them into a single class of problems whose solution was based on procedures that were in a certain sense homogeneous. Thus the path was open toward establishing the mutual connection of the two principal operations of the future infinitesimal analysis. It is probably not coincidental that the initiator of this replacement of the method of exhaustion by the method of infinitesimals was an astronomer, an opticist, and at the same time a mathematician: Johannes Kepler.

The investigations of previous centuries had gone by, again from the point of view of official mathematics, mostly in the geometric framework; this is to a significant extent the case also in the seventeenth century. However at the same time the problems of the future differential and integral calculus took on a more and more analytic garb. The history of the establishment of one of the basic theorems of integral calculus is quite indicative in this regard.

Cavalieri, an exponent of the geometric algebra of the ancients, in obtaining the theorem
$$\int_0^a x^n\, dx = \frac{a^{n+1}}{n+1},$$
was obliged to resort to multidimensional geometry, which did not yet exist at the time, and to apply multidimensional, or, as Cavalieri expressed it, "imaginary" solids, although he recognized the difficulty of such expressions (especially for the first half of the seventeenth century), noting at the same time that in the case $n > 3$ it is easier to work with purely algebraic methods [1, pp. 287–303]. In contrast Fermat, Roberval, and some others obtain these propositions analytically, and indeed more simply and briefly (*History of Mathematics...* [1, pp. 183–187]).

To apply the method of exhaustion to obtain such a general result and *a fortiori* to use this method to find the extremum of an expression defined by an analytic formula of a rather complicated form—which mathematicians of the first half of the seventeenth century had also studied—was possible only after it had been transformed into the method of limits. But this was still very far off. For that reason the rejection of the method of exhaustion and its replacement by the method of infinitesimals were unavoidable as a consequence of the greater generality of the phenomena undergoing investigation, which were now represented in mainly analytic form.

In this period there was one important difference between the concepts of derivative and integral. One can say that in the seventeenth century a notion of definite integral was created that was to have a long stay in mathematics: the notion of the integral as the area bounded by a curve, the axis of abscissas, and two ordinates, approximated by inscribed and circumscribed step figures whose accuracy increased according to the fineness of the partition. We encounter this notion in Valerio as early as 1604 (Bortolotti [2, pp. 38–39]), then in more transparent form in Mengoli in 1659 (Agostini [1]), then in almost its modern form in Newton in 1686 [5, pp. 29–30]. The mathematicians of the seventeenth century managed to reduce all problems of finding areas, volumes, centers of gravity, and lengths of curves to this concept.

There was no similar general notion of the derivative as the slope of the tangent to a curve encompassing all the problems of differential character at the time (Zeuthen [1, p. 316]). In solving such problems, however, the notion of a certain general infinitesimal operation arose that made it possible to find the tangents to curves, velocities and accelerations, and the maxima and minima of analytical expressions, and that had a rather clearly expressed algorithmic character. Fermat, Roberval, Wallis, Gregory, Barrow, and many others essentially managed to solve all the differentiation problems of the time and did so without any particular difficulty, using the fundamental rules of the differential calculus, not, however, distinguishing them, especially as algorithmic rules.[2] The latter was done by Leibniz [1].

Before the seventeenth century the integral had been regarded as a certain number associated with an object being studied that remained invariant in the course of the investigation (more precisely, as the ratio of two numbers, one known and one unknown). The time had now come to change this approach. This change took place outside the realm of mathematics, in the study of mechanical motion. In 1638 Galileo defined uniformly accelerated linear motion and established the rule that, in modern notation, can be stated by the formula $v = gt$, where v is the velocity of the body, t the time of the motion, and g a constant factor. He then proved that the path traversed by the body up to the instant t is expressed—again in modern notation—in the form

$$x = \frac{1}{2}gt^2,$$

i.e., by the indefinite integral

$$\int_0^t v(t)\,dt = \int_0^t gt\,dt = \frac{1}{2}gt^2.$$

---

[2] For details cf. *History of Mathematics...* [2, pp. 192–196].

And this is the first example of an indefinite integral in the history of science (Bortolotti [2, p. 124]).

These arguments of Galileo were generalized by Torricelli, who considered more and more general ways in which velocity could depend on time. In this case also he arrived at the result that when the velocity of the motion is given as a rather general function of time, the distance traversed in time t is expressed by the integral $\int_0^t v(t)\,dt$ (Bortolotti [2, pp. 124–126]).

Thus the problem of measuring a new mechanical quantity—the distance traversed by a body—led to a modification of the integral that soon became the dominant one and made it possible to establish a connection between the derivative and the integral. Along with this went the idea that the derivative was anterior to the integral: the distance was determined by the given velocity.

Fermat had conjectured this connection (Bourbaki [3, p. 227]); Torricelli evidently recognized it (Bortolotti [1, pp. 150–152]). It was rather profoundly studied by Gregory and Barrow, who actually established the mutually inverse character of the concepts of derivative and indefinite integral (*History of Mathematics...* [2, pp. 210–214], Scriba [1, pp. 32–39], Zeuthen [1, pp. 351–356]). Nevertheless neither Gregory nor Barrow became the creator of the differential and integral calculus, first of all because their notions were too geometric or kinematical and second because they were insufficiently algorithmic.

One circumstance involving Barrow's theorem on the mutually inverse character of the connection between the derivative and the indefinite integral deserves special attention. We have said that a general notion of integral arose in the seventeenth century, but that there was no general notion of derivative. It was this historical course of development of analysis, where the methods of finding integrals were developed in more detail, that evidently caused Barrow to make the strategic error of giving integration precedence over differentiation.

> Barrow's inversion theorem makes it possible, having found the result of a differentiation or integration, to obtain from it the result of applying the inverse operation to the result of the original operation. Up to that time integration in the form of quadrature was the better known of the two operations. For that reason Barrow was inclined rather to deduce a method of finding tangents from the known quadratures than to do the reverse (Zeuthen [1, p. 355]).

This partly closed off from him the road to an algorithm of infinitesimal analysis, but at the same time, in this "error" of Barrow one can see the germ of the modern tendency to invert the relation between the concepts of derivative and integral that characterized classical analysis.

4.3 The analysis of Newton and Leibniz

The idea of mathematizing human knowledge is one of the most seductive ideas in the history of science; in one form or another it has accompanied the history of science at least since the time of Pythagoras and down to the present, assuming the most diverse forms. As early as the time of Ancient Greece the mathematization of mechanics, astronomy, and optics was such that, according to Bochner, "the Greeks themselves wondered whether, and in what sense, these [sciences] could be distinguished from mathematics proper" [1, p. 179]. This idea was vigorous also in the Middle Ages (*History of Mathematics...* [1, pp. 269, 273–283]); and in the first half of the seventeenth century Galileo, along with some general declarations (Zubov [1, p. 209]) began the concrete realization of this idea following the Greeks. In general this century in the history of scientific thought was so tightly bound up with mathematics that Dannemann had grounds for saying that at the time "mathematics and mathematical physics, together with philosophy, newly freed from the fetters of scholasticism, came to be identified with science in general" [1, p. 199]. It is superfluous here to talk about the whole connection of mathematics with science in the seventeenth century. We shall pause to discuss only one side of this connection, which in our view determined its character for a very long period of time.

As Kolmogorov has emphasized,[3] the fundamental idea that guided Newton in his scientific activity was that of the mathematization of natural science, and the scientific analysis that he created was one of its principal realizations. This mathematization assumed first of all that the totality of notions in mechanics about the basic concepts as the fundamental science of the time had been formed. And indeed the fundamental object of mechanics—the study of motion with its basic attributes—was in many respects completed by the time of Newton. Probably the most essential aspect of the notion of movement that had been created was its differential character, i.e., movement was thought of as taking place in space from point to point by passing through all the intermediate states that succeed one another in a continuous manner during a continuous flow of time (Kuznetsov [1, pp. 30, 64–65, 160]). To describe such a motion it was insufficient to know only the initial and final

---
[3] In an introductory speech at a reception dedicated to the 250th anniversary of the death of Leibniz held 1 December 1966 in Moscow. Cf. also Kolmogorov [3, p. 27, footnote 2].

positions of the moving body, as was the case in Aristotelian physics; one had to be able to take account of the position of the object at every instant of time. The integral characteristics of the motion—distance as a function of time, a finite velocity as a function of acceleration, and the like—entered as the results of completed differential motions, as something derivative and secondary in relation to the motion taking place. And these integral characteristics themselves turned out to be not fixed, but varied as time went on; they depended on the differential characteristics.

At least two essential points that occurred in the development of analysis up to the time of Newton contradicted the notion of mechanical motion just described. First, integration was long considered to be a primary operation, for the most part independent of the operation of differentiation. Second, the integral itself was thought of as a definite integral, in the form of a fixed number assigned to the invariant quantity in question. To mathematize the idea of motion that was taking shape, it would be necessary to reverse the relation of these things: The operation of differentiation would have to come to the fore, and integration would have to be regarded as derived from it; the integral itself would have to be changed from definite to indefinite.

But to carry out such considerations, which incidentally were not really recognized by anyone in such a general formulation at the time, such ideas as the primacy of differentiation in relation to integration and the primacy of the indefinite integral in relation to the definite integral were insufficient by themselves. A new mathematical apparatus was needed, one suitable for describing the mechanical phenomena being studied. This apparatus was to a significant degree prepared by the preceding development of algebra.

It is in algebra that the notion of a variable quantity arises. "The turning point in mathematics was Descartes' *variable magnitude*. With that came *motion* and hence *dialectics* in mathematics, and *at once also of necessity the differential and integral calculus*" (Engels [1, p. 199]). The form of generality that was born in algebraic symbolism, made it possible to think of the content this symbolism described as changing and to represent it as flowing and moving.

Further, although the main goal of algebra at the time was considered to be the solution of equations, algebra was also already being regarded as the study of the general properties of algebraic operations. But the operations in algebra had mutually inverse characters. In passing to infinitesimal operations, it was natural to regard them in this way.

Algebra also made it possible to go beyond visual geometric representations, within whose framework the majority of the previous results of analysis had been obtained. In algebra the geometric image was replaced by a comparatively simple formula that was visualizable in a different sense and on which transformations could be performed according to established rules of an algorithmic type. Moreover because of the fact that letters playing the role of parameters can occur in algebraic expressions, the solution of an algebraic problem frequently appeared as the solution of an entire class, sometimes even an infinite class, of problems of a similar type.

The conversion of the operation of differentiation into the primary operation of analysis, the replacement of the definite integral by the indefinite integral, the algebraization of the basic concepts of the analysis of the time, the creation of the algorithms of infinite series and differential calculus, the compilation of tables containing primitives and derivatives of functions, the application of the new calculus in manifold different problems of mathematics and mathematical science—such were the problems confronting Newton and Leibniz that were solved by them and their successors.

We do not propose to describe the creation of analysis. There is a voluminous historico-scientific literature on this question, and it will be reasonable to mention only a few sources: Zeuthen [1, pp. 374–425], *History of Mathematics...* [2, pp. 215–287], Boyer [1], Baron [1], Pogrebysskii [2, pp. 212–268]. For us what is important is to note certain features of this analysis. In the infinitesimal analysis created by Newton, Leibniz, and their successors the following properties were clearly noticeable:

a) the operation of differentiation is considered to be an operation that could be carried out for any analytical expression considered at the time; special cases of it are the geometric and mechanical methods of finding tangents, extrema, and the like;[4]

b) differentiation is proposed as the basic, primary operation of analysis, and integration enters merely as its inverse; in this connection the concept of a definite integral, which had earlier been the basic object of study, fades into the background, into the area of applications, and not only was there no special symbol for it for quite some time, there wasn't even any special name for it; the value of a definite integral is obtained from the formula

$$\int_a^b f'(x)\,dx = f(b) - f(a),$$

which reduces it to the difference of two particular values of the primitive;[5]

---

[4] Provided, of course, one is considering the factual results obtained in analysis; the situation is more complicated in regard to the foundation of analysis.

[5] The Newton-Leibniz formula. It was stated explicitly by Newton, in geometric disguise, however [2, p. 89]; it seems not to occur in explicit form in the writings of Leibniz, but it follows easily from his approach to integration as the inverse of differentiation and his explicit mention of the need for an additive constant.

c) the transition from the views of differentiation and integration as methods of finding geometric or mechanical quantities to the notion of these operations as the differentiation and integration of functions defined by analytic expressions is completed.

d) starting with Newton [2], more and more extensive tables of primitive and derived functions are complied; various analytic techniques are developed for finding primitives; new transcendental functions are introduced and studied; the view arises of the integral calculus as the study of transcendental functions that arise in the course of finding primitives.[6]

It is interesting that these characteristics of analysis in the narrow sense, i.e., as the study of differential and integral calculus, defined it as a certain open system, whose problems could occupy the minds of many and many a generation of analysts. It would suffice merely to keep on introducing new transcendental functions, studying their properties and creating more complicated differential expressions using them, developing the technique of analytic computation, etc. Such an approach was to a large degree carried out in the history of analysis; it can be said that it continues to some extent even today. On this route some remarkable achievements were attained. It suffices to mention elliptic functions, the integral logarithm, and the like, which were interesting not only intrinsically, but also for their applications in algebra, the theory of numbers, geometry, natural science, and technology.

At the same time with such views of differentiation and integration and their relation to each other mathematical analysis would have been a closed system in the sense that the concepts of function, derivative, and integral, would not have been enriched in their theoretical content, being confined within certain bounds. This did not occur principally because of a new separation of these two concepts—a repetition in a sense of the situation that existed in antiquity and in the Middle Ages. This separation itself had, in the final analysis, mathematical causes.

---

[6] Such a view is found in Euler [3, Vol. I, p. 61]; but it was evidently most clearly voiced by Gauss in 1808: "In the integral calculus I find much less interesting the parts that involve only substitutions, transformations, and the like, in short, the parts that involve the known skillfully applied mechanics of reducing integrals to algebraic, logarithmic, and circular functions, than I find the careful and profound study of transcendental functions that cannot be reduced to these functions. We now know how to handle circular and logarithmic functions as well as we know that one times one is one, but the magnificent gold mine in which the treasures of the higher functions are buried remains as yet almost *terra incognita*. I have worked on this a great deal in the past, and will some day publish my own large opus about it." (This promise was not kept. Cited in Markushevich [2, p. 170]).

4.4 The groundwork for separating the concepts of derivative and integral

In the nineteenth century the development of the concepts of derivative and integral begins to diverge more and more from the path described in the preceding section, and again, as in antiquity and in the Middle Ages, the leading concept is that of the integral, which had grown too large for the procrustean bed of classical analysis.

The causes leading to the creation of an independent concept of definite integral were many.

First of all, the problems that were solved by calculating definite integrals had never left the agenda of mathematical research. On the contrary, the circle of problems requiring such computations for their solution had significantly expanded, especially because of astronomy, mechanics, and physics. And in mathematics itself definite integrals had begun to be used to calculate the sums of infinite series; certain transcendental functions had begun to be expressed in terms of them; they were being used to represent the solutions of differential equations, and so forth. Thus the concept of a definite integral had begun to play an ever more visible role, and an ever increasing number of investigations were being devoted to the independent study of it unconnected with the notion of integral as the difference of two values of a primitive. It suffices to mention that Euler's work on definite integrals alone occupies two volumes of his *Opera omnia*. The fact that such investigations, especially in multidimensional cases, were mostly conducted outside the limits of the concept of integration as the inverse of differentiation can be explained as follows.

The theoretically beautiful method of obtaining the value of a definite integral by the Newton-Leibniz formula soon exhausted its possibilities even in the one-dimensional case, so that Luzin could write with some justice in 1933 that "for 150 years after the death of Euler mathematicians were unable to make any breach in the ring of integrations he had forged" [8, p. 363].

Euler himself, perhaps the most ardent exponent of the view of integration as the inverse of differentiation, recognized to some extent the limitations of this view. Noting that in the cases where a primitive cannot be found "we have no choice but to try to find a value [of the integral—F. M.] as close as possible to the true value" [3, Vol. I, p. 61] and passing to the approximate computation of integrals [3, Vol. I, pp. 200–229], he really introduces the notion of an integral as a limit of sums.[7]

---

[7] For more details cf. *History of Mathematics...* [3, pp. 345–346]. In essence all of the exponents of the view of integration as the difference of values of a primitive, sensed its limitation;

The situation was even worse in the two- and three-dimensional cases. Up to 1910 mathematics contained no concept corresponding to the physical notion of density, tension, etc. for area and volume analogous to the concept of a derivative for a function of one variable, except for the prefigurations of Cauchy [3] and Peano [1]. In a certain sense the concept of a total differential, which replaced it in relation to differentiation and integration, was unsatisfactory, because examples of fairly simple differential expressions that were not total differentials had long been known, and consequently there could be no question of inverting them. But in mechanics, astronomy, physics, and even geometry it was necessary to compute more two-dimensional and three-dimensional quantities of integral type than one-dimensional. These quantities had appeared as early as the time of Newton in the form of multiple integrals rather than iterated integrals, which were required for the concept of integration as the inverse of differentiation (Antropova [1]). But the reduction of multiple integrals to iterated integrals is a ticklish business, and multiple integrals were arising predominantly in the form of definite integrals as limits of certain sums (Fikhtengol'ts [2], *History of Mathematics...* [3, pp. 349–352]).

Thus, alongside the official, so to speak, conception of integration as the inverse of differentiation, the integral was all the time showing through it in the form of a certain type of sum, which was either completely unconnected with traditional differentiation or possibly connected with it in some unknown way.

There was yet another essential factor promoting the downfall of the notion of the definite integral as the difference of two values of a primitive. In 1768 d'Alembert [1] discovered that the expression of the definite integral by the Newton-Leibniz formula was unsuitable when the integrand becomes infinite on the interval of integration, and he illustrated this point with examples from mechanics and geometry. Lagrange, Gauss and Poisson later came to this same conclusion. But in the traditional derivation of the Newton-Leibniz formula no restrictions of finiteness were imposed formally on the derivative: by definition
$$\int f'(x)\,dx = f(x) + C;$$
assuming that the integral vanishes at $x = a$, we have $C = -f(a)$, and then for $x = b$ we have
$$\int_a^b f'(x)\,dx = f(b) - f(a).$$

---

limitation; this is attested to by the approximate methods of calculating integrals, as well as the preservation of the old tradition of interpreting the integral as a certain sum.

4.4 Separating the concepts of derivative and integral

Consequently the question was now not only about the insufficient applicability of this formula in practice, but also about a serious theoretical defect in it. It is therefore not coincidental that both Gauss and Poisson incline explicitly toward the notion of the integral as a certain sum.[8]

Existence problems, which had occupied an ever larger place in mathematical research since the beginning of the nineteenth century, played a role in the change of the relations between the definite and the indefinite integrals. Referring to the article of Yushkevich [2, pp. 386–389], we shall discuss in detail only the possible refraction of these questions in the concept of an integral.

Some time after such great achievements as finding the integrals of rational functions and certain classes of irrational and transcendental functions, the notion of integration as the inverse of differentiation began to encounter difficulties of an analytic nature. The method of introducing new transcendental functions, which seemed applicable from a theoretical point of view, led to extreme difficulties in practice, even in the case of integrals of the form $\int f(x, \sqrt{R(x)})\,dx$, where R is a polynomial of degree three or four and f is a rational function. For a long time it remained unclear how many and what kind of transcendental functions it was necessary to introduce in order to integrate such expressions. Only at the end of the eighteenth century did Legendre succeed in reducing all the various elliptic integrals to three types; but in order to do so he had to write a forbidding two-volume treatise.[9] This treatise was merely the point of departure for a gigantic subsequent labor on the part of a large number of nineteenth-century analysts in the study of elliptic functions. It hardly seemed possible to take this route in more complicated cases.

But then a question of the following type naturally arose. Suppose it is required to find a specific $\int f(x)\,dx$ for a comparatively complicated $f(x)$. The computation of such an indefinite integral sometimes required herculean labor, even when $f(x)$ was not particularly complicated, and sometimes it simply couldn't be done. And when existence questions appeared on the mathematical agenda, it was legitimate to ask whether it was reasonable to try to compute such an integral, not only exactly, but even approximately, if it was unknown whether $f(x)$ even belonged to the class of functions for which a primitive exists. A noncircular proof of the existence of a primitive for a continuous function turned out to be very difficult, and was achieved only at the beginning of the twentieth century by Lebesgue [11]. In the meantime it was much simpler to prove the existence of a definite integral for a rather

---

[8] For more details cf. Yushkevich [2, pp. 397–399].
[9] Cf. *History of Mathematics...* [3, pp. 354–360].

large class of functions.

Essential though the circumstances just noted were, individually and collectively, for the appearance of new conceptions of integration and differentiation, nevertheless in our view, as in the transition to classical analysis, a more important factor was the atmosphere in which the mathematization of physics was being conducted at the beginning of the nineteenth century. One could hardly expect to construct a connected chain of logical inferences leading from the general idea of the mathematization of physics to the relatively particular conclusion that the concept of a definite integral should be taken as the primary one in integral calculus. Nevertheless an approximate chain of this type can be observed.

By the end of the eighteenth century mathematical analysis had grown and penetrated physics to such an extent that for many scholars it was not only a mathematical discipline, but a universal physical theory (Pogrebysskii [1, p. 16]). But in order to serve as such a universal theory, it had to be significantly enlarged. Indeed, although the concept of motion, as before, constituted the main object of investigation in mathematical science, although the differential character of motion noted in the last section continued to be accepted tacitly as the basis of its nature from the mathematical point of view, nevertheless the content of this concept had been significantly enriched and deepened, and the analytic methods at the disposal of the eighteenth century were no longer adequate for an understanding of it. This applied primarily to two important types of motion—various oscillatory processes and the questions of heat propagation. In studying motions of this type it turned out that to describe them by mathematical means it made sense to apply an entirely new method of representing functions analytically—trigonometric series and integrals—since the functions encountered in physical problems were turning out to be of a more complicated character—nonanalytic and even discontinuous.

Such a mathematical apparatus had begun to appear even in the eighteenth century,[10] but credit for creating it belongs to Fourier. His *Théorie analytique de chaleur* was a turning point in the development of mathematics and mathematical physics. This book has already been, and will obviously continue to be, the subject of numerous historico-mathematical studies. At present we are interested only in two aspects of it connected with the topic under discussion.

The first of these is the essential elevation of the role of the definite integral in physico-mathematical investigations, which is reflected in this book. One of the main problems in expanding functions in trigonometric series was

---

[10] Cf. Paplauskas [1, pp. 7–38].

4.4 Separating the concepts of derivative and integral

to determine the coefficients of the expansions. Fourier solved this problem, and in doing so found that the most reasonable technique for calculating the coefficients was by use of a definite integral. Thus the problem of expanding a function in a trigonometric series was connected with the concept of a definite integral, and this connection has manifested itself fruitfully down to the present.[11] The role of the definite integral in studying functional dependence was elevated even further in connection with the introduction of the Fourier integral, since with it the integral itself entered as the method of representing a function analytically.[12]

We shall give a somewhat more detailed discussion of the second aspect.

Since, as we have said, in the seventeenth and eighteenth centuries motion was thought of as having a differential character, it was natural that the principal analytic technique for studying it became (and remains to a large degree even today) differential equations. However, in the nineteenth century a new analytical technique—integral equations—came to be applied more and more widely to the study of physical facts. We are not so bold as to claim that some sort of conceptual revolution had taken place in the views of the nature of motion, leading to the replacement of one mathematical method of describing reality by another in many cases. It is a fact, however, that the application of integral equations in the study of natural phenomena was very widespread, that integral equations were occupying an ever larger place in purely mathematical investigations, and that the study of differential equations was more and more frequently being reduced to the study of integral equations. In addition doubts were frequently expressed as to the legitimacy of the differential description of the phenomena of the outside world.[13] And in our view it is a very interesting circumstance that the creation of the theory of integral equations took place before the definite integral attained the status of a concept independent of differentiation in Cauchy's lectures [2] of 1823.

In speaking of the appearance of integral equations in science it is customary to begin with a note of Abel in 1823.[14] But here is what we can read

---

[11] In historical perspective it is unimportant that other mathematicians, Euler in particular, had worked on this connection even in the eighteenth century (Paplauskas [1, pp. 32–38]). It was only after the appearance of Fourier's book [1] that it became widespread. Fourier himself arrived at it independently in a more general and rigorous form; he pointed out this independence in 1829 [2, p. 174].

[12] For the Fourier integral cf. Paplauskas [1, pp. 236–256].

[13] We refer, for example, to Lebesgue's statements [20, pp. 291–292], to the numerous papers of Gyunter, of which we note only his article [2], as being most typical in this regard, and to the favorable evaluation of the latter's position by Smirnov and Sobolev [1, pp. 12–13].

[14] Cf., for example, Hahn [1, p. 72].

in the book of Fourier in 1822:

> If one is considering the expression
> $$\int_{-\infty}^{+\infty} \varphi(\alpha)\,d\alpha \int_{-\infty}^{+\infty} \cos p(x-\alpha)\,dp,$$
> it is clear that it is a certain function of x... The nature of this function will obviously depend on the nature of the function chosen as $\varphi(\alpha)$. One may pose the question as to what kind of function $\varphi(\alpha)$ must be in order to obtain a given function $f(x)$ after the two integrations have been carried out. In general finding the integrals intended to represent physical phenomena reduces to problems like the preceding. These questions have as their object *to determine arbitrary integrands of a definite integral in such a way that the result of the integration is a given function* [our emphasis—F. M.] [1, p. 474].

Fourier investigates this integral equation further and shows [1, pp. 474–488] that the integrals of a series of boundary-value problems of mathematical physics reduce to solving this equation. Consequently we see here not only the statement and solution of a new and important mathematical problem and a clear recognition of its connection with the problems of mathematical physics, but also the clearly expressed idea of reducing the solution of differential equations to the solution of integral equations. Of course from the point of view of later requirements one frequently cannot consider Fourier's arguments rigorous, but almost no mathematical argument looks perfectly irreproachable a short time after it has been carried out, or sometimes even at the time it is carried out when looked at by a person with a different mindset.

Thus the growth of the role of the definite integral, the fact that in some cases it could not be reduced to the classical definition, the posing of existence questions, the observation of a transition to the integral description of reality—all this dictated the need for a new view of the fundamental infinitesimal operations. In addition, by the end of the first quarter of the nineteenth century the mathematical foundation for a new approach to these operations was in place: the general concept of a function of a real variable had matured (Euler, Lacroix, Fourier); the class of continuous functions, which was fundamental for the nineteenth century, had been distinguished (Bolzano, Cauchy); a systematic study of the apparatus of trigonometric series as a means of representing a functional dependence analytically had begun (Fourier, Poisson, Cauchy, and others); techniques of reasoning based on the concept of a limit had been developed (mainly by Cauchy)—to mention only the purely analytic conditions. The earlier view of the definite integral as a particular thing, the simple difference of two special values of a primitive, was in contradiction to the fundamental role that the concept of a definite integral was beginning to play in both mathematics itself and the mathematical description of reality.

4.5 The separation of differentiation and integration

The transition to a new conception of integration was prepared at the end of the eighteenth and beginning of the nineteenth centuries, and a certain part was played in this preparation by Lacroix, Lagrange, Gauss, and Poisson.[15] However, their achievements are minimized in at least two respects. First, they all had in mind the integration of analytic functions, and second, when they studied a definite integral, it was nevertheless a secondary concept for them in relation to the indefinite integral.

It was Cauchy [2] who carried out this transition in 1823, reversing the relation between the indefinite and definite integrals and applying it to a significantly larger class of functions. He starts from the concept of a continuous function [2, p. 11] and introduces the definite integral for it as the limit of sums [2, pp. 111–117], proving that this limit exists. The indefinite integral is characterized as a definite integral but with a variable upper limit [2, p. 141], i.e., on the logical level the former is in some sense secondary in relation to the latter.

On the level just described a complete break between differentiation and integration had not yet occurred: differentiation of the indefinite integral of a continuous function leads back to the integrand at each point, a primitive exists for every continuous function, and the Newton-Leibniz formula holds (Cauchy [2, pp. 140–145]). But the separation of these two operations from each other is already noticeable: the fundamental theorems of integral calculus are proved mostly on the basis of the notion of a definite integral as a limit of sums, and not by referring to the inversion of the corresponding theorems of differential calculus, as had been the case in the preceding period.

Moreover, in this book [2] Cauchy proposed yet another concept of the integral, which he did not connect, indeed could not connect, with any kind of differentiation at all, and this is a fact of equal interest in the present context. What is involved is the integral in the sense of a principal value, defined by the following formula, for example, for a function $f(x)$ that is continuous and bounded on (a,b) except for a single point ξ with $a < \xi < b$ in a neighborhood

---

[15] On this see the paper of Yushkevich [2].

of which it increases without bound,

$$\int_a^b f(x)\,dx = \lim_{\varepsilon \to 0} \left(\int_a^{\xi-\varepsilon} f(x)\,dx + \int_{\xi+\varepsilon}^b f(x)\,dx \right),$$

if this limit exists (Cauchy [2, pp. 133–134]). This integral generally received no recognition during the nineteenth century, although it was occasionally studied. It was only after Hardy [1–3] rehabilitated this concept at the beginning of the twentieth century by establishing that it has many of the "good" properties of the ordinary integral and can be successfully applied in various problems that it began to be used more and more widely in the most diverse problems, especially in the theory of trigonometric series. But however the nineteenth-century mathematicians felt about this concept of integral, another disquieting symptom for the connection between differentiation and integration could be seen clearly.

A retreat from this connection began from the side of differentiation. The latter had previously been considered a simpler and more direct operation that could always be performed as long as analytic functions were being discussed; it was known how to calculate the derivative for any function proposed.

The introduction of the general concept of a continuous function complicated the situation greatly. As we have said (p. 65), continuity and differentiability were separated from each other, and the suspicion even arose that there exist continuous but nondifferentiable functions. At the same time every continuous function is integrable in the Cauchy sense, i.e., in relation to one and the same class of functions the operations of differentiation and integration behaved very differently, and no necessary connection between them could be seen. This circumstance was fully recognized by Dirichlet at the beginning of the 1850's. In his lectures at the University of Berlin[16] after noting that in the general case it is impossible to prove the existence of the derivative an arbitrary continuous function and even asserting that "it is possible to define in a purely graphical manner functions that have a derivative nowhere" [3, p. 20], but at the same time conceding that every continuous function is integrable, he makes the following radical inference:

> Because of this essential difference in the character of the integral and the derivative, and because, as we have seen, the concept of a definite integral can be obtained independently of the differential

---

[16] The following discussion concerns his lectures of the summer of 1854, which were published by Arendt only in 1904. This edition, according to Klein [1, p. 99], is closer to Dirichlet's train of thought than the earlier (1871) edition of Meyer.

calculus, contrary to the usual custom of regarding the integral calculus as its inverse, there is much to be said in favor of giving an independent basis to the two concepts of integral and derivative, especially since many difficulties whose resolution involves a great deal of trouble, can then easily be overcome. [3, p. 20].

In relation to the definite integral Dirichlet carried out such a program in his lectures. He not only obtained the basic properties of the integral by regarding it as the limit of sums, but also calculated, for example, the integral $\int_a^b x^k\,dx$ directly from the definition, without using the simpler technique that follows from the Newton-Leibniz formula. Moreover, like Barrow, he solved certain problems of differential calculus starting from the concept of a definite integral. Thus, noting that the proof of the existence of a derivative for an exponential or logarithmic function is very difficult in the context of differential calculus without resorting to series expansions, Dirichlet shows that such a proof can, in contrast, be obtained very simply using the definite integral [3, pp. 20–21].

Although Dirichlet's lectures remained unpublished for a long time, his ideas about differentiation and integration were noted even outside mathematics proper. As an illustration of this last assertion we note the brochure of the engineer L. L. Mazurkiewicz [1], published in 1875 in Petersburg, which in the mathematical sense is rather confused, not to use a stronger expression, but is characterized by the successive implementation of the idea of regarding integration as an operation independent of differentiation, even to the point of a direct computation of the integrals of the fundamental elementary functions.

But the realization of Dirichlet's programs also began very quickly in mathematics proper. In December of 1853 Riemann presented to the Philosophical Faculty of Göttingen University the manuscript of his paper "Über die Darstellbarkeit einer Funktion durch eine trigonometrische Reihe" [2][17] in which the concept of a definite integral was introduced, studied, and applied with essentially no connection to differentiation.

There was one essential difference between the integrals of Cauchy and Riemann. While the Cauchy integral turned out to be capable of redefinition as the difference of values of a primitive function (though not until long after its introduction) since the existence of a unique primitive for every continuous function was proved by a method independent of the concept of the

---

[17] The fact that Dirichlet's lectures, published in his book [3] were read in 1854 does not contradict what has been said: he could have read similar lectures previously, and if he did, Riemann would have had the opportunity to hear them during the time he was in Berlin in the years 1847–1849.

integral (Lebesgue [11]), this turned out to be immeasurably more complicated for the Riemann integral. With the Riemann integral mathematicians had entered the world of discontinuous functions, where they encountered many unexpected things.

There had been equally many surprises in the region of continuous functions, which have already been noted—the absence of derivatives, or, if one may so express it, the presence of "primitive" functions without corresponding derivatives; for a definite integral defined as the difference of values of a primitive, there was no integrand; for the "distance" described by such a continuous motion no velocity existed; a curve defined by such a function had no tangent or length, and so forth. With the transition to discontinuous functions the number of such "nuisances" increased sharply. Dirichlet's example [1, p. 23] gave a function having no primitive, since this function belonged to the second Baire class, while a derivative is always of Baire class 1 at the highest. Differentiation of the indefinite Riemann integral led to the integrand only at points of continuity of the latter (Darboux [1, p. 76]); in other words the indefinite integral was not necessarily a pointwise primitive. In 1881 Volterra [2, pp. 17–20] constructed an example of a continuous function having a bounded derivative that is not Riemann integrable; in other words, the primitive function and its derivative explicitly existed, but the integral, regarded as a limit of Riemann sums, was not adequate to establish the connection between them, although for this case the definition of the integral as a difference of values of the primitive was adequate.

Thus the nature of the Riemann integral itself and the functions that were integrable (or nonintegrable) using it now dictated the need for studying the properties of the integral that are not necessarily connected with differentiation: integrability conditions (Riemann [2, pp. 240–241], Hankel [2, pp. 28–29], Darboux [1, pp. 64–73], Smith [1, pp. 141–144], Ascoli [1, pp. 869–871], Thomae [1, pp. 11–12], and others); a different form for the definition of the Riemann integral, not as a limit of sums, but in terms of the identity of the upper and lower integrals (Ascoli [1, p. 867], Jordan [2, pp. 439–440], Peano [3], and others); the connections of the theory of the integral and the theory of measure (the papers of Hankel, Smith, and Jordan already mentioned and others) and so forth.

Mathematicians set out on a similar route in the study of differentiation. As early as 1841 Ostrogradskii had introduced the symmetric derivative [1, pp. 166–167];[18] later this derivative was studied by Du Bois-Reymond [4, pp. 122–123]; Riemann applied a generalized second derivative [2, p. 246]; these derivatives were studied and applied by many mathematicians, but there was

---

[18] Cf. Remez [1, p. 58, footnote].

4.5 The separation of differentiation and integration

usually no question of any kind of inverse for them, and when such an inverse was discussed, the discussion was rather unintelligible, as for example in Du Bois-Reymond [6]. In 1875 Thomae generalized the ordinary derivative to right- and left-hand derivatives[19] and even proved the important theorem for integral calculus that if one of the derivatives of $F(x)$ is zero everywhere on an interval, then $F(x)$ is constant on that interval. This theorem enabled him to generalize the concept of a primitive for a function $f(x)$ as a function $F(x)$ whose right (or left) derivative equals $f(x)$ [1, p. 8]. However, he speaks only in passing of a definite integral that can be introduced in terms of one of these derivatives [1, p. 9].

The next major step in the study of differentiation was taken by Dini, who introduced the derivates [4, §145]. There were now four "derivatives" corresponding to a continuous function, and what was needed for a connection between differentiation and integration was an independent study of the relationship of these derivates, the differences among them, conditions under which they coincide, and so forth. A large part of Dini's book is devoted to this study.[20] The matter became still more complicated when derivates began to be studied in general.

Perhaps the most important step in the separation of integration from differentiation during the nineteenth century was the introduction of the Stieltjes integral. To be sure, the first hints of this new concept of integral in the writings of Cauchy [3] and Peano [1, Ch. V] were connected with a new type of derivative,[21] but in the nineteenth century they remained only hints, not very clearly stated and not appreciated by other mathematicians. But in 1894, when Stieltjes [1, pp. 67–68] approached this new concept of integral more rigorously, and in 1897, when König [1] did the same, this concept was no longer connected with any kind of differentiation; it then remained in this state with Voronoi, Lyapunov, and other scholars, with the exception of Markov, until 1916, when W. H. Young [5] combined Stieltjes integration and the differentiation of one function with respect to another. Incidentally, there is an interesting detail here also: Galois had stated the idea of differentiating one function with respect to another in 1830 and even attempted to prove—though it is difficult to say to what extent he recognized this—the existence of such a derivative for "every" function.

Thus the more profound study of the fundamental infinitesimal opera-

---

[19] We are not claiming that he was the first to introduce these derivatives. Timchenko [1, pp. 490–491] actually points out that they can be found in the writing of Arbogast; it is likely that they existed even earlier.

[20] For some details, cf. Hawkins [1, pp. 48–50].

[21] For details cf. Medvedev [3].

tions led more and more toward a separation of them. This does not mean that there was beginning to be a conscious rejection of the connection between them in the nineteenth century. The separation of these operations was, it seems, merely an objective tendency, by no means recognized by all the mathematicians who were studying them, especially those who were applying them in specific investigations. They were subjectively striving rather to preserve the earlier relations between differentiation and integration. When the two operations were discussed, they were for the most part interpreted in the spirit of classical analysis, not to mention the fact that in the overwhelming number of textbooks on analysis at the time integration was as a rule treated as a sort of inverse of differentiation, and the definite integral was for the most part regarded, even in works of pure research, as the difference of values of a primitive. Even within the limits of a few examples, one can mention the following.

This is the approach to the relation between differentiation and integration of continuous functions taken by Lobachevskii [1, p. 42] in 1834, and this is eleven years after the introduction of the Cauchy integral, although it is difficult to suppose that he was not aware of Cauchy's book [2], which had been translated into Russian in 1831.

In 1875 Thomae [1], who did much to develop the theory of the Riemann integral, still preserves both definitions of this integral—as the difference of values of a primitive and as a limit of sums—and tries to reconcile them, although he himself asserts [1, p. 15] that it is difficult to prove the existence of a primitive without using the definition of the integral as a limit of sums.

Here is what Volterra wrote in 1887:

> Differentiation and integration are regarded in analysis as two infinitesimal operations; they are defined independently of each other and *it is proved that each operation is the inverse of the other* [emphasis ours—F. M.] [3, p. 209].

And this is after 1881, when he himself had proved [2] the existence of bounded derivatives that are not Riemann integrable, i.e., the impossibility of connecting the classical differentiation and integration of the nineteenth century! This can be explained only by the fact that he regarded the definition of the integral as the difference of values of a primitive as appropriate in 1887 (and later).

The number of such examples could be greatly increased. We shall mention only one other cycle of investigations related to the generalization of the Riemann integral to unbounded functions and connected with the names of Dirichlet, Du Bois-Reymond, Cantor, Scheffer, Harnack, Hölder, Stolz, Schönflies, and Vallée-Poussin. The investigations named here have been studied by Pesin [1, pp. 21–39] and Hawkins [1, pp. 71–79], and there is no need to go into detail about them. We remark only that they were to a large extent aimed at resurrecting in a certain sense the notion of the integral as the difference of values of a primitive. But such a notion required the existence of a unique continuous function $F(x)$ satisfying the equality

$$F(x'') - F(x') = \int_{x'}^{x''} F'(x)\,dx$$

for every closed interval $[x', x''] \subset [a, b]$ on which $F(x)$ is continuous. In the meantime during the 1880's continuous nonconstant functions were discovered that were constant on the intervals complementary to an arbitrary nowhere dense set. In this connection numerous investigations were needed at the end of the nineteenth century to produce the concepts of Borel measure, absolutely continuous function, the Denjoy integral, etc. Thus although the separation of the two fundamental infinitesimal operations had essentially already occurred, the idea of a connection between them, which had given rise to classical analysis, remained (and in many respects still remains at present) alive and active.

Volterra's result on the existence of a bounded derivative whose pointwise primitive cannot be recovered using the Riemann integral delivered perhaps the most telling blow to the connection of differentiation and integration. And when it subsequently became possible to restore this connection using new integral constructions, this was always regarded as a great mathematical achievement. Thus Lebesgue, after introducing his integral and showing that every bounded derivative was integrable in his sense, wrote with justifiable pride in 1902:

> The integral of a bounded derivative, regarded as a function of the upper limit of integration, is a primitive function of the given derivative. Consequently the basic problem of integral calculus is theoretically solved once and for all, once the given derivative is bounded [3, p. 233].

The title of the second of Denjoy's papers [1, 2] is also revealing: "Calcul de la primitive de la fonction derivée la plus générale." In this 1912 paper he writes with no less pride than Lebesgue:

> Thus the problem inverse to differentiation can be solved for the first time in all its generality: knowing that f is a derivative, find a method of calculating its primitive... [2, p. 1077].

The method was provided by the integral he had introduced.

These were indeed great achievements of mathematical thought. They were insufficient, however, in certain respects. We shall describe this in somewhat more detail.

The starting point for classical analysis was, as has been stated more than once, the fact that differentiation was adopted as the primary operation and integration was regarded as a mere inverse of it, without the application of any summation process; the definite integral was simply the difference of values of the function obtained by such an inversion. It was not always possible to implement this view systematically, but theoretically it prevailed and was regarded as the only correct one. After Lebesgue proved in 1905 that there exists a primitive for every continuous function, the Cauchy integral in the form of a limit of sums became practically unnecessary. The situation was different for the Lebesgue and Denjoy integrals at the time they were introduced. For them it was precisely the summation-type construction that was essential, and significant efforts of Vitali, Lebesgue, Luzin, Denjoy, and Riesz were required in order to do in relation to the Lebesgue and Denjoy integrals what Lebesgue had done for the Cauchy integral—get rid of the summation-type construction in their definitions, which in addition was complicated by transfinite reasoning in the construction of the Denjoy integral. To do this it was necessary to introduce the concepts of an absolutely continuous function and a function of generalized bounded variation and study their properties.[22]

The approach of Perron [1] was more radical in this respect. In his construction all summation-type constructions were absent, and integration entered merely as the inverse of differentiation in pure form.[23] It would seem that the classical conception had triumphed at this point. However the situation was much more complicated.

The first complication was that these constructions related only to ordinary differentiation. But, as has already been pointed out in part, other differentiations existed along with it, whose numbers were increasing; thus in 1912 Borel proposed a special definition of derivative, as did Khinchin in 1915.[24] There was as yet no corresponding inverse for them.

The second circumstance was even more vital. If the basic type of integration in the first decade of the twentieth century had been integration with respect to the independent variable, from 1909 on, after Riesz' result on the expression of a linear function on the space of continuous functions in the form of a Stieltjes integral was established, the integration of one function with respect to another occupied ever firmer positions. For this integral, as

---
[22] We shall not dwell on this, referring to the book of Pesin [1, pp. 148–176].
[23] Ibid.
[24] Cf. Pesin [1, pp. 177–178].

4.5 The separation of differentiation and integration

we have said, there was at first no corresponding differentiation at all, so that there was nothing to invert, and from the introduction of such a differential operation to its inversion to the original function in relation to generalized infinitesimal operations took more than a decade and a half. The theories of integration of Stieltjes type in the form of the Lebesgue-Stieltjes integral, the Denjoy-Stieltjes integral and the Perron-Stieltjes integral developed very gradually.

The situation became even more complicated in the transition from the one-dimensional case to the multi-dimensional case, and still more when the domain of abstract spaces was entered. Fréchet described the latter situation eloquently in 1928, and we shall use his words:

> The situation is different for abstract differentiation and integration.[25] These two concepts lie in opposite relations to the topology of abstract spaces.
>
> Let me make this more precise. Radon has obtained an interesting definition of a Euclidean integral while giving a suitable generalization of the concept of the Stieltjes and Lebesgue integrals. Carrying his generalizations to their conclusion, inspired by the methods of Young, we have arrived at a definition of the integral of a functional on an abstract space.[26] This can be done independently of the concept of a neighborhood introduced in this abstract space.
>
> In contrast the concepts of neighborhoods, which are not needed to define an integral, become necessary when we attack abstract differentiation, but not sufficient. It is not enough that we can say when the increment is small, i.e., one position of the variable is close to another. It is also necessary that there be a possibility of separating the differential of y from the variable y, i.e., of understanding what it means to say that the result of subtraction is infinitely small compared to the increment of the abstract variable x.
>
> It seems difficult to give any meaning to this condition without assuming that the spaces, whether different or not, in which x and y vary have not only a topological character but also a vector character (however, these two properties must naturally be connected with each other) [8, p. 272].

---

[25] Fréchet had just been discussing the introduction of the concept of a continuous operator in a functional space.

[26] Fréchet gives a reference to his paper [5] at this point.

It is true that the Fréchet integral was so subsumed by the Lebesgue integral in the 1930's that it is frequently called by the latter name, thereby connecting it with the correspondingly generalized differentiation. However, it was followed by the integrals of Burkill, Kolmogorov, and Glivenko for real-valued functions in abstract spaces, which have not to this day been defined by any differential process. The situation was still further complicated in the transition to vector-valued functions, when the integrals of Graves, Bochner, Birkhoff, Dunford, Gel'fand, Gavurin, Pettis, Phillips, Price, Rickart, Ionescu-Tulcea, and others were proposed. Along with the various types of integrals various types of abstract differentiation were introduced and studied, for the most part independently of any integrations and without any connection to them.[27]

Thus if we speak on the whole of a connection between differentiation and integration in the twentieth century, it must be clearly stated that the Newton-Leibniz-Euler conception of the primacy of differentiation and of integration as the inverse of differentiation, which was realizable for a comparatively small class of functions of one real variable, turns out on the whole to be very far from being realized, despite the enormous efforts and great achievements in this cause. This was bound to be reflected in the search for new approaches, one of which we shall try to trace in some detail in the following sections.

It is clear that the establishment of connections between differentiation and integration was not the only problem of interest to the scholars who undertook the study of these operations. There were many other problems, and we shall attempt to give an idea of only one of them.

So many different generalizations and modifications of the concepts of derivative and integral have been proposed in the twentieth century that one cannot help recalling the situation that arose in the seventeenth century before the revolution brought about by Newton and Leibniz. At that time quadratures and cubatures of numerous areas and volumes and the lengths of various curves were calculated, the locations of the centers of gravity of various figures were determined, the tangents to various curves were found, and the extrema of many analytic expressions were investigated. More or less general techniques were developed for solving large classes of problems and two classes of these techniques were distinguished—differential and integral. The establishment of connections between various problems belonging to the same or different classes was one of the most important mathematical problems of the time.

Something analogous can be observed at the present time. Only now

---

[27] For a survey of the latter cf. Averbuch and Smolyanov [1, 2].

4.5 The separation of differentiation and integration

the role of individual quadratures, rectifications, computations of tangents, and the like is played by the numerous concepts of integral and derivative. To establish connections between them, again both within the class of derivatives and the class of integrals and between representatives of the different classes, to determine whether one concept is subordinate to another or to find the common domain of two distinct concepts is also an important problem of the theory of functions today.

We shall illustrate this last assertion with several examples.

At the beginning of the twentieth century the Lebesgue integral made a triumphal procession through mathematics. Along with it, but less conspicuously, rather in the shadow of applications, the concept of the Stieltjes integral was developing. One of the current problems of mathematics at the beginning of the century was that of finding an analytic expression for a linear functional on the space of continuous functions. To solve it the Riemann integral was applied in 1903 (Hadamard [1]) and the Lebesgue integral in 1904 (Fréchet [1]), but a satisfactory solution was found only in 1909 by Riesz [6], using the Stieltjes integral. In the following year Lebesgue [15] made a bold attempt to get by without the latter in this problem, reducing it to his integral. However in 1914 W. H. Young [5] established that the Stieltjes integration procedures, when suitably generalized, include as a small particular case both the Riemann and the Lebesgue procedures. And though Lebesgue (and some others as well) nevertheless later attempted to carry out the idea of [15], it was in essence an idea that could not be correctly carried out.

In 1914 Perron [14] proposed his own definition of the integral. He himself did not yet know that his concept was broader than that of Lebesgue; he showed only that it is not narrower, but at the same time is simpler in certain respects. After that a cycle of papers was required to clarify the mutual relation of this integral with the existing concepts of integral and the restricted Denjoy integral; we mention, for example, the papers of Bauer [1], Hake [1], Aleksandrov [1, 2], and Looman [1]. The introduction of the Khinchin-Denjoy integrals and the generalization of the Denjoy and Khinchin-Denjoy integrals to the corresponding integrals of Stieltjes type suggested corresponding modifications of the original definition of Perron, and a large number of papers was also devoted to this, an enumeration of which would occupy a great deal of space. It was not only such large integration processes that required clarification of their interrelationships. Denjoy alone proposed no less than ten different definitions of the concept of the integral, and the total number of such definitions is evidently over a hundred. For that reason establishing connections among them and clarifying their interrelationships grew into a huge mathematical problem, seemingly far from a satisfactory solution. And if earlier the successive generalizations were to some extent contained in one another[28]—the integrals of Cauchy, Riemann, Lebesgue, Denjoy, Khinchin-Denjoy and the Perron integrals equivalent to the last two, nowadays constructions appear more and more frequently that are not imbedded in one another, such as the integrals of Kolmogorov for finite and countable partitions, the A-integral and the restricted Denjoy integral, etc. How untidy a job it is to clarify the relationships between the various types of integrals even for functions of one variable is quite clear from the survey of Vinogradova and Skvortsov [1].

A similar picture appears for differentiation. And though Khinchin [1] succeeded to some extent in "imposing" a linear ordering in relation to the various definitions of derivative in 1924,[29] this ordering was more and more violated as time went on. And now the words of Khinchin sound even more ominous:

> The various generalizations of the concept of a derivative were created independently of one another and not only does the region of applicability of each of them remain unexplored, even the relative power of the various methods of differentiation has not been subjected to a systematic study [1, pp. 377–378].

4.6 The Radon-Nikodým Theorem

Extending his one-dimensional theory of integration to the multidimensional case in 1910, Lebesgue proved the following theorem.

Let $F(e)$ be an additive and absolutely continuous function of the L-measurable set e having a finite, determinate derivative $f(x)$ almost everywhere. Then

$$F(e) = \int_e \psi(x)\, d\mu, \qquad (1)$$

where $\psi(x)$ equals $f(x)$ at points where $f(x)$ is finite and assumes arbitrary finite values at the remaining points, $\mu(e)$ is the Lebesgue measure of the set (e), the derivative $f(x)$ is obtained by differentiating $F(e)$ with respect to μ over a regular family of sets contracting to the point at which the derivative is sought, and the sets e are taken in n-dimensional Euclidean space [16, p. 399].

This theorem is one of the generalizations of the Newton-Leibniz theorem to the case of the multiple Lebesgue integral, and in the one-dimensional

---

[28] Although there were also deviations from the "linear ordering" type.
[29] Actually in 1919; Khinchin's paper [1] is dated 1919, but was not published until 1924.

case it corresponds to the theorem that an absolutely continuous function of a point is the indefinite integral of its derivative. In this theorem the role of the difference $F(b) - F(a)$ in the Newton-Leibniz theorem is played by the value of the function F on the set e, and the role of the derivative $f(x)$ is played by the derivative of F over the regular family of measurable sets.

Three years later Radon, in generalizing the Lebesgue theory of integration to the Lebesgue-Stieltjes theory, proved [1, p. 1342–1351] a theorem that is analogous in certain respects.

If $F(e)$ is completely additive and $\mu(e)$ now denotes a Lebesgue-Stieltjes measure, then

$$F(e) = \int_e f(x)\, d\mu, \qquad (2)$$

where $f(x)$ is a certain point function that is measurable with respect to μ.

In contrast to Lebesgue Radon did not develop a theory of differentiation parallel to the integration he introduced, and for that reason his formula (2) could not be regarded as a further generalization of the Newton-Leibniz theorem. However, the analogy between the formulas (1) and (2) is sufficiently transparent (at least nowadays) so that the function $f(x)$ can be regarded as a kind of "derivative" of the function $F(e)$ in the second case also. This $f(x)$ is constructed from $F(e)$ in a certain manner—rather complicated in Radon's paper—and the process of constructing $f(x)$ from $F(e)$ can be taken as the definition of the operation of differentiation. Radon did not take this step, most likely because of the complexity involved in carrying it out. To do so it would have been necessary in addition to overcome a complicated psychological barrier. Indeed, the tradition that continues down to the present of regarding differentiation as a simpler direct operation was in conflict here with the method of obtaining $f(x)$ from $F(e)$. It was not primary in the case under consideration; on the contrary, the intrinsically rather complicated operation of integration was the primary one, and $f(x)$ was determined, again in a very complicated manner, from a certain "inversion" of the integral. Another cycle of papers was required before the $f(x)$ of formula (2) began to be called a derivative.

In 1915 Fréchet [5] generalized Radon's reasoning still further. His main results can be briefly described as follows.

Lebesgue and Radon had considered sets in n-dimensional Euclidean space and functions defined on such sets. Lebesgue had limited himself to sets that were measurable in his sense, and Radon had generalized Lebesgue measure to Lebesgue-Stieltjes measure. Both had introduced these measures using a certain process that amounted to constructing a measure by enclosing sets in intervals, determining the measures of the intervals enclosing the set in the elementary-geometric sense or in the sense of Stieltjes, and finding the infimum of the latter. All of their subsequent constructions were based on this process.

Fréchet wished to introduce an integral for real-valued functions defined on abstract sets (for functionals). With such great generality it was not possible to introduce a measure analogous to that of Lebesgue, and Fréchet found an ingenious way out of this situation.

Fréchet introduced the concept of an additive family (class) T of sets, i.e., a family of subsets E of some abstract space X such that

1) if $E_i \in T$, then $\cup E_i \in T$ for any finite or countable sequence of sets E_i;

2) if $E_1 \in T$ and $E_2 \in T$, then $E_1 \setminus E_2 \in T$,[30] and Fréchet declared every set of the space X belonging to the given family T to be measurable with respect to this family, not giving any process for constructing a measure.

Relying on this definition of measurability for abstract sets, he was able to construct a theory of integration for point functions defined on sets that are measurable in this sense.[31]

Fréchet's theory of integration, like that of Radon, was not accompanied by a corresponding operation of differentiation. Moreover the analogue of formula (2) was missing in it. This last gap was filled in 1930 by Nikodým [1] in 1930, when he proved that if $F(e)$ is a completely additive set function on an additive class T on which the measure μ is defined, then

$$F(e) = \int_e f \, d\mu, \qquad (3)$$

where f is a point function measurable with respect to μ [1, pp. 168–179]. This theorem has come to be named the Radon-Nikodým Theorem in the mathematical literature. It differs from Lebesgue's theorem, like the analogous theorem of Radon, in that the point function f is not obtained by any preliminary differentiation, but is constructed from the set function F.

---

[30] Fréchet did not state the requirement that the empty set belong to T; it followed from (2). A slightly different definition of an additive class can be found in Saks [2, p. 7]. Fréchet's formulation corresponds to the definition of a σ-ring (Halmos [1, p. 24]).

[31] We remark that it was the Fréchet integral, methodologically refined, that was considered in the book of Saks [2]; in that book it is called the Lebesgue integral. It is also of interest to note that Radon actually introduced an additive class using similar axioms, but only for sets in n-dimensional spaces [1, p. 1299]. However, he did not use them to introduce measure, but defined the measure constructively.

The theorems of Lebesgue, Radon, and Nikodým, especially the last of these, has been the object of a very large number of studies continuing down to the present. We shall mention only a few papers that are interesting in several respects.

W. H. Young [6] in 1916 and Daniell [1] in 1918, not knowing of Radon's paper [1], introduced an operation of differentiating one function with respect to another for the one-dimensional case and connected it with the Lebesgue-Stieltjes integral. In particular both of them converted Radon's formula (2) into an analogue of the Newton-Leibniz theorem, since for them the function $F(x)$ in this formula played the role of the derivative in the earlier, generalized sense and was obtained using a preliminary definite differentiation independent of any integration process. For Euclidean spaces of higher dimension the problem turned out to be more delicate, since Vitali's covering theorem was involved [4]. Daniell [3, 4], by modifying in various ways the definition of the derivative of one function with respect to another, was able to obtain the analogues of formula (2) with an $f(x)$ that was a derivative in these senses, without using Vitali's theorem. Some time later Maeda [1] established the same result by suitably generalizing Vitali's theorem.

The situation regarding differentiation in the generality involved in Nikodým's formula (3) turned out to be much more complicated. Here we merely note the works of Hayes and Pauc [1] and remark that under certain assumptions of a rather complicated nature similar results were obtained for formula (3) also. The situation becomes even more tangled when vector-valued measures are discussed.

Thus the interpretation of formula (3) and its generalizations to vector-valued measures turned out to be a very messy matter when looked at from the point of view of a preparatory differentiation. Mathematicians then made a "knight's move"; instead of first introducing the concept of a derivative and then connecting it with the integral, they made the concept of integral the primary one and proved that formula (3) holds for "some" function $f(x)$, and then called that $f(x)$ the derivative of the set function $F(e)$ with respect to the set function $\mu(e)$. With such a definition of the derivative $f = dF/d\mu$ many of the usual theorems of integral calculus can be obtained (Halmos [1, pp. 128–132]). It is interesting that this leap of thought was made not in developing the Fréchet theory that is involved in formula (3), but in studying more general types of integrals, although the new definition of the derivative was actually introduced for the Fréchet integral.

4.7 The relation between differentiation and integration in the works of Kolmogorov

After the introduction of the Fréchet integral the next significant step in the theory of integration was the transition from the integration of a point function to the integration of a set function, and in connection with it the transition to a new construction of integral sums. Burkill [1] was the first to carry out this construction, which in brief consisted of the following:

In all the preceding definitions of the integral as the limit of sums or the common value of the bounds on the sums the integral sums were formed from products of the values of a point function (or the supremum or infimum of its values on a set) and the values of some set function (measure) on a certain subset of the domain of definition of the function, i.e., products of the form $f(x)\mu(e)$ or $M_i\mu(e)$, where M_i is the supremum or infimum of $f(x)$ on e. Heterogeneous elements—a point function and a set function—were thereby combined in a sense in the construction of the integral sums, and integration itself was thought of precisely as integration of a point function with respect to a measure.

Burkill broke with this tradition. In his constructions the point function plays no role at all in the beginning: it is a set function (specifically an interval function for Burkill) that is integrated, and the integral sums are formed from the values of this set function on the partial subsets into which its domain of definition is partitioned.

The second essential step in Burkill's approach consisted of extending his reasoning to nonadditive set functions.

Burkill also considered differentiation of interval functions and some of its connections with his integration. However, in the aspect we are considering this is not particularly interesting, and we shall leave it aside. Burkill's arguments were later developed by many mathematicians, and he himself used and enriched them many times.[32]

Kolmogorov raised the theory of integration to a new level of generality in 1930. His paper "A study of the concept of an integral" was a very broad synthesis of the abstract ideas of Fréchet and Moore and the fertile ideas of Burkill regarding integration of nonadditive set functions.

After noting in the introduction that there are two basic approaches to the concept of integral—the attempt to solve the problem of the primitive for a hypothetical derivative and the desire to give some general summation process—Kolmogorov conjectured that there were fundamental difficulties in

---

[32] A survey, albeit incomplete, of the papers in this area up to 1948 can be found in the paper of Ringenberg [1]. Many papers on the Burkill integral also appeared after 1948.

the attempts to state a concept of integral general enough to encompass both of these approaches to the problem of integration[33] [2, p. 655]. Therefore, renouncing any such attempts, he set himself the goal of stating the most general concept of limit-of-sum type encompassing all similar concepts, confining the discussion to real-valued functions, though he mentioned the possibility of extending his arguments to vector-valued functions [2, pp. 658–659].

In contrast to Fréchet, who started with a σ-ring of sets, Kolmogorov started with a multiplicative domain of abstract sets, which he introduced and which is defined as follows:

Let abstract sets be given.[34] A system of sets possessing the property that the intersection of any two sets of the system is a set of the same system is called a *multiplicative domain* or simply an $\mathfrak{M}$-*domain* [2, p. 661]. The abandonment of the operation of abstract set-theoretic union in favor of the operation of intersection was evidently dictated by two considerations. First, when two types of additivity—finite and countable—were present, it was not clear which should be regarded as anterior to the other. As we have said, Fréchet and the majority of mathematicians after him tended to prefer countable additivity. In the meantime Kolmogorov showed that two essentially different integration theories were obtained, neither contained in the other, according as finite or countable additivity was chosen. Second, in defining the integrals he was considering Kolmogorov wanted to use the generalized Moore-Smith limit [1], or, as it is more frequently called nowadays, the limit over a filter. The most important component of the latter is that, given two elements of a set, it is possible to find in the set within which the limit is sought a third element that in some sense follows the two given elements. The operation of set-theoretic intersection made it possible to carry out such a choice: it suffices to take the intersection of the two first elements and regard this as an element that follows both of them.

Suppose given an $\mathfrak{M}$-domain of abstract sets. Any decomposition of a set $E \in \mathfrak{M}$ into a finite or countable number of pairwise disjoint summands belonging to $\mathfrak{M}$, i.e.,

$$E = \bigcup_n E_n, \quad E_n \in \mathfrak{M},$$

is denoted DE.

---

[33] The half-century that has passed since this conjecture was stated has not only confirmed it, but has failed almost completely to reveal the nature of the difficulty, although there has evidently been no lack of attempts at combining these approaches.

[34] Nowadays the words "in some abstract space X" are added. Neither Fréchet nor Kolmogorov did this, though in practice they proceeded in this way.

Let $E \in \mathfrak{M}$, and let the function f, in general not single-valued and nonadditive, be defined for elements of the decomposition DE. Taking into account Burkill's idea on the nature of the integral sums and generalizing it, Kolmogorov forms his integral sum in the form

$$\sum_n f(E_n). \qquad (1)$$

The integral of f over the set E with respect to $\mathfrak{M}$ is defined as the generalized limit of the sums (1) and is denoted

$$(\mathfrak{M}) \int_E f(dE). \qquad (2)$$

Depending on whether countable or finite decompositions D are chosen two types of integrals (2) are obtained. Here Kolmogorov discovered, seemingly for the first time in an investigation of the concept of integral, that one and the same function may be integrable in both senses of (2), yet the values of its integrals over the same region of integration might not agree, i.e., the integrals with countable partitions and finite partitions conflict with each other [2, p. 698].[35]

Kolmogorov devoted most of his paper [2] to developing the conception of integration with no connection to differentiation. He succeeded in constructing extremely general theories of this type. But the idea of a connection of the two infinitesimal operations is too seductive, and Kolmogorov did not avoid it. It was he who made the leap of thought mentioned at the end of the preceding section, reversing the relation between differentiation and integration that had predominated in classical analysis. To this topic he devoted a special appendix to [2] with the title "On the foundation of differentiation on abstract sets" [2, pp. 694, 696].

It is interesting that he made this reversal not in relation to his own integrals, but for the Fréchet integral, and even then not in full generality. To be specific Kolmogorov used formula (3) of Sec. 6, calling the point function that is the integrand the derivative of one set function with respect to another and proving its existence and uniqueness under certain assumptions. He proved the existence of a derivative by defining a certain constructive process "which in significantly more general cases can be regarded as the definition of the

---

[35] Similar facts were later discovered for other integrals also. Thus, for example, Vinogradova [1] established that the indefinite A-integral in general conflicts with improper Lebesgue integral and the restricted Denjoy integral.

derivative itself" [2, p. 695]. It is evident that in introducing the concept of derivative Kolmogorov had resorted to the Fréchet integral rather than his own integrals (2) because point functions and set functions enter the former as independent elements from the very beginning of the construction of the theory, while Kolmogorov's theories are based exclusively on the consideration of set functions. Point functions arose in his theory in order to show that the conception he had developed encompassed the preceding ones, and for this demonstration he proposed a technique of reducing a point function to a set function. The derivative of one function with respect to another is always a point function and by its nature it is in some sense anomalous in the context of his basic constructions.

It is likely that differentiation is not introduced in the theory of integrals generalizing the Kolmogorov integrals for the same reasons. Of the latter we may mention, for example, the theories of Glivenko [1, 2] and Ionescu-Tulcea [1].[36]

4.8 The relation between differentiation and integration in the works of Carathéodory

In a cycle of papers that began to appear in 1938 and are summarized in [4], which appeared posthumously in 1956, Carathéodory developed an idiosyncratic theory, in the context of which the relations between differentiation and integration acquired some new features. In what follows we shall rely on the book [4], tracing the problem we are interested in only in its most general features.

Carathéodory starts from an abstract Boolean algebra,[37] whose elements he calls *somas*. These somas are for him the materials out of which he constructs his theories of measure and integration, which to limit ourselves to to a very general and approximate description, are as follows.

The measure and integral introduced by Carathéodory are essentially those introduced by Fréchet in 1915, and which, as we have said, are expounded in the book of Saks [2] under the name of Lebesgue measure and Lebesgue integration. The only difference is that Carathéodory constructs his theories not for sets defined in abstract spaces and point functions of such sets, but for elements of an abstract Boolean algebra and real homomorphisms of the σ-field of all Borel sets of real numbers into that Boolean algebra,[38]

---

[36] The latter is essentially Glivenko's theory extended to set functions with values in abstract spaces. We shall not discuss the other definitions of Ionescu-Tulcea.

[37] For Boolean algebras cf., for example, Sikorski [1].

[38] In regard to real homomorphisms cf. ibid., pp. 328–329.

called *place functions* (Ortsfunktionen) by Carathéodory and serving as the analogues of point functions.

This transition from abstract sets to the elements of an abstract Boolean algebra was a new step in the direction of generality in questions of measure and integration, so that in this direction Carathéodory went further than Kolmogorov. However, while Kolmogorov was developing his ideas for set functions, Carathéodory confined himself to integrating place functions, and in this respect he did not achieve the generality of Kolmogorov.

It seems to us that the main peculiarity of the integration theory developed by Carathéodory is the following. In Fréchet integration a set function and a point function are introduced as two independent concepts. But Carathéodory bases everything on the soma function—the Boolean analogue of a set function—and from it he constructs the corresponding analogue of a point function. His reasoning runs as follows [4, pp. 72–76].

He fixes a certain soma M and expands it into a finite number of pairwise disjoint somas M_i:

$$M = \sum_{i=1}^{p} M_i. \qquad (1)$$

To each of the M_i he assigns a real number y_i, i.e., he introduces the simplest function of an element of the Boolean algebra. If we now introduce the symbol f to describe the fact that a definite y_i corresponds to each M_i, we obtain what Carathéodory calls a finite-valued place function. In the case of ordinary sets f is merely the step function defined on M by the equalities $f = y_i$ on M_i. Under different partitions (1) of the soma M and choice of the various numbers y_i different finite-valued place functions are obtained. Carathéodory shows that the set of finite-valued place functions is a lattice. This lattice is incomplete, and, completing it in the usual way, he calls the elements of the complete lattice so obtained *generalized place functions* defined on the soma M. He also gives a direct construction of such generalized place functions [4, pp. 76–83] that is a generalization of the well-known description in the theory of functions for a point function $f(x)$ defined on a set E using its Lebesgue sets $E(y_k \leq f < y_{k+1})$.

After studying the real homomorphisms (Ch. IV) and measures and Boolean algebras (Ch. V) thus introduced, Carathéodory introduces and studies the concept of the integral of a homomorphism with respect to a measure (Ch. VI). We shall not discuss this in detail, but turn to the question of the connection between integration and differentiation that we are interested in.

Like Kolmogorov, but in the more general context of a Boolean algebra

4.8 Differentiation and integration in the works of Carathéodory

and in more detail, Carathéodory begins with the Radon-Nikodým Theorem

$$F(e) = \int_e f \, d\mu. \tag{2}$$

Relying on it, he shows [4, pp. 191–193] that, first, if the functions f_1 and f_2 are nonnegative and integrable, i.e., the integrals

$$F_1(e) = \int_e f_1 \, d\mu, \quad F_2(e) = \int_e f_2 \, d\mu,$$

exist, then the integral of the product $f_1 f_2$ with respect to μ exists and

$$\int_e f_1 f_2 \, d\mu = \int_e f_2 \, dF_1 = \int_e f_1 \, dF_2, \tag{3}$$

and, second, if the measurable function f is finite and positive, then each of the equalities

$$F(e) = \int_e f \, d\mu, \quad \mu(e) = \int_e \frac{1}{f} \, dF \tag{4}$$

follows from the other. Then remarking that the equality

$$F(e) = \int_e dF$$

follows from the definition of the integral, he deduces [4, p. 193] that the first of equalities (4) can be written in the form

$$\int_e dF = \int_e f \, d\mu$$

or, discarding the integral signs,

$$dF = f \, d\mu.$$

Thus an object arises for Carathéodory that he calls the abstract differential. Equalities (3) and (4) make it possible to introduce a formal calculus for these differentials resembling the ordinary calculus of differentials in analysis and the theory of functions. Thus, from the differential equalities $dF = f_2 \, d\mu$ and $d\mu = f_1 \, d\nu$ by means of (3) one obtains the equality $dF = f_1 f_2 \, d\nu$; from the equalities $dF = f\,d\mu$, and $d\mu = \dfrac{1}{f}\,dF$ it follows from (4) that each of these can be obtained from the other; combining these last results gives the majority of the formulas of the usual calculus of differentials.

Thus, starting from functions of elements of a Boolean algebra and real homomorphisms, Carathéodory introduced formally a very general calculus of differentials. However he did not venture to introduce the derivative of one soma function with respect to another in this generality, limiting himself to the remark [4, p. 193] that in order actually to obtain the operation of differentiation, i.e., for some kind of constructive way of finding a real homomorphism of two measures, it was necessary to apply a special construction that he had used earlier [4, pp. 185–188]. And while Kolmogorov, in obtaining a point function from two set functions, says frankly that the process itself can be taken as the definition of the derivative, Carathéodory [4] does not make such a radical inference, although the construction he mentions can also be regarded in this way. Moreover, when he makes his general considerations specific, restricting them to the theory of Lebesgue integration in n-dimensional space, he takes a step backwards, introducing differentiation and integration independently of each other and only afterwards establishing a connection between them [4, pp. 304–309].

4.9 A few more general remarks

The results of Kolmogorov and Carathéodory that have been briefly described here were of course not the final links in the chain of development of the idea of the inverse relationship between the operations of differentiation and integration. This idea continues to be developed and enriched. But we have already gone far beyond the bounds of the theory we delineated, which consists of functions defined on sets in n-dimensional spaces and assuming real values; further advance in this direction would require an even greater detour into abstract algebra, functional analysis, and topology.

Meanwhile the problem of the relation between the operations of differentiation and integration is quite complicated and many-faceted even in the theory of functions in the narrow sense and has not been fully explored mathematically, so that a historical analysis of it would be premature. For that reason we shall confine ourselves to some general remarks.

We have already said that so many different operations of differentiation and integration have been invented that the problem of establishing connections among them is very complicated. As for any common point of view on this, it simply doesn't exist; and even in specific situations the picture is rather pessimistic. It suffices, for example, to point out the following fact.

4.9 A few more general remarks

The classical indefinite integrals considered earlier always belonged to comparatively small classes of continuous functions. The A-integral, which has been energetically studied recently can be defined (Vinogradova [1]) so that every measurable function on $[a, b]$ coincides almost everywhere with a function $F(x)$ that is an indefinite A-integral existing almost everywhere. When this is done, there is an entire family of functions possessing the same indefinite integral, but differing on a set of positive measure. In other words, even if we introduce some operation making it possible to "differentiate" this indefinite integral, the differentiation cannot be single-valued and must lead to essentially different functions depending on the way in which it is carried out. "Thus there is no differential connection between the indefinite A-integral and the integrand" (Vinogradova and Skvortsov [1, p. 82]).

This last inference is mostly based on the idea of the independent introduction of the operations of differentiation and integration and the subsequent establishment of a connection between them. But it is also of interest to study fully the situations in which one starts with one of these operations and defines the other in terms of it. The route from the derivative to the integral—the route of classical analysis—has been very widely studied, and many remarkable results have been obtained on it, but a satisfactory explanation of the essence of the relation between the two infinitesimal operations has not been obtained on this route. The independent introduction and subsequent connection of the two operations—the route of the theory of functions of a real variable—has also not yet yielded positive results in general. Up to now the approach based on the primacy of integration and the introduction of differentiation in terms of it has been less studied, and the works discussed in the preceding sections, to which many others could be added, is only one branch of the complete stream of investigations in which mathematicians start from the concept of an integral in defining the derivative. We shall say a few more words about another branch.

In 1950 S. L. Sobolev [1] proposed a new definition of derivative. As its basis he chose not the concept of integral itself, as Kolmogorov and Carathéodory had done, but the formula for integration by parts for integrable point functions. His definition of a generalized derivative can be formulated as follows.[39]

Let D be a bounded region in n-dimensional Euclidean space, defined by the points $x = (x_1, x_2, \ldots, x_n)$. We shall say that a function $\psi(x)$ is *of compact support* in D if it equals zero outside some region D' contained strictly in the interior of D.

---

[39] In what follows we shall give it in the formulation of Smirnov [1, pp. 338–339], which differs only in insignificant details from that of Sobolev.

Assume that the functions $\varphi(x)$ and $\psi(x)$ have continuous derivatives up to order l inside D and the function $\psi(x)$ is of compact support. Considering some derivative of order l

$$D^l\varphi = \frac{\partial^l\varphi}{\partial x_1^{l_1}\partial x_2^{l_2}\ldots\partial x_n^{l_n}},$$

applying the formula for integration by parts, and taking account of the compact support of $\psi(x)$, we obtain the equality

$$\int_D D^l\varphi(x)\psi(x)\,dx = (-1)^l \int_D \varphi(x)D^l\psi(x)\,dx. \tag{1}$$

The equality (1) is made the basis of the definition of the concept of a derivative, and the idea of the generalization arises from the fact that this equality can hold for functions $\varphi(x)$ that are nondifferentiable in the usual sense.

Let the functions $\varphi(x)$ and $X(x)$ be integrable over any region D' strictly interior to the region D, and suppose that for any l times continuously differentiable function $\psi(x)$ of compact support the relation

$$\int_D X(x)\psi(x)\,dx = (-1)^l \int_D \varphi(x)D^l\psi(x)\,dx \tag{2}$$

holds. Then the function $X(x)$ is called the generalized derivative of the function $\varphi(x)$ in the region D.

Sobolev's definition suffers in comparison with those of Kolmogorov and Carathéodory where abstractness is concerned. On the other hand Sobolev's theory of generalized derivatives is very profoundly developed and has found numerous and important applications. We cannot discuss this in detail, and we mention only the vast cycle of papers on imbedding theorems that mostly arose from the Sobolev definition of a generalized derivative.[40] We also mention the interesting circumstance (in our view) that a connection has recently been found between the Kolmogorov-Carathéodory approaches and that of Sobolev, the latter being subordinate to the former.[41]

The definition of the generalized Nikol'skii derivative (Nikol'skii [1, p. 214; 2, p. 168]) is also closely connected with the Sobolev definition. Nikol'skii's definition is also essentially based on the concept of integral. And a connection has been found between this definition and a definition of Kolmogorov-Carathéodory type similar to the one just noted for the Sobolev definition.[42]

---

[40] Cf. Kudryavtsev and Nikol'skii [1].
[41] Cf. Klement'ev and Bokk [1].
[42] Cf. Klement'ev and Bokk [2].

This ramification confirms yet again the interest in Kolmogorov's idea of the primacy of the operation of integration and the definition of the operation of differentiation in terms of it. The idea is not yet sufficiently developed to be the leading idea of infinitesimal arguments, but seems to be steadily penetrating into quite diverse investigations and not only those of function-theoretic nature.

Chapter 5
Nondifferentiable continuous functions

5.1 Some introductory remarks

The problem of nondifferentiable continuous functions has been studied by a very large number of mathematicians and historians of mathematics. Of the papers in which the history of this problem has been traced in more or less detail one can, for example, mention the following: Pascal [1, pp. 91–128], Brunschvicg [1, pp. 337–340], Pasch [3, pp. 122–129], and Hawkins [1, pp. 42–54]. These historical excursions are far from complete, understandably so due to the difficulty of the problem. We do not propose to give anything like a complete description of its history. Our goal will be to trace the general features of a single idea: the change in the view of mathematicians on the relationship between a function and its derivative from certainty as to the existence of a derivative for every function to the establishment of the fact that the set of functions having an ordinary derivative at even one point is in a sense negligibly small in comparison to the whole set of continuous functions, that the main bulk of continuous functions, so to speak, consists of functions not having a derivative at any point. A few more detailed remarks will be made, but only in passing.

To give an idea of the complexity of the problem, let us consider a few of the simplest cases.

Suppose a function $f(x)$ is given on $[a, b]$. We ask ourselves the question, Does there exist another function $\varphi(x)$, $x \in [a, b]$, such that there exists some connection of differential character making it possible to obtain the values of $\varphi(x)$ from those of $f(x)$?

A subquestion immediately arises as to the nature of differential connections, since there are many such connections. For example, this connection might be expressed as the limit of the difference quotient

$$\lim_{h \to 0} \frac{f(x+h) - f(x)}{h} = \varphi(x) \qquad (1)$$

as h varies continuously over all real numbers (the usual derivative). The variation of h can also be restricted by the requirement that it assume only positive or only negative real values (right- and left-hand derivatives).

A differential connection of the form (1) need not be defined using the ordinary limit; the upper and lower limits are frequently taken instead, and the result is either two derivatives (upper and lower) or the four Dini derivates (upper right, lower right, upper left, and lower left).

Going further and varying h in (1) not over all real numbers, but only over countable sequences of them, one obtains simply derived numbers. And one can go still further.

Every relation of the type (1) in the circumstances described makes it possible to talk about a certain differential connection between $f(x)$ and $\varphi(x)$.

We now give another example. Instead of (1) we can consider the limit of the difference quotient

$$\lim_{h \to 0} \frac{f(x+h) - f(x-h)}{2h} = \varphi(x) \qquad (2)$$

and repeat all the arguments given above in relation to it. Then (2) will give another class of differential relations between $f(x)$ and $\varphi(x)$.

Yet another example occurs if (1) and (2) are replaced by

$$\lim_{k \to 0} \lim_{h \to 0} \frac{f(x+h+k) - f(x+h) - f(x+k) + f(x)}{hk} = \varphi(x).$$

In each of these examples it is a question of particular types of differential connections between the functions $f(x)$ and $\varphi(x)$, and in the case when $\varphi(x)$ exists for a given $f(x)$, the function $\varphi(x)$ is the "derivative" in some sense or other of the function $f(x)$. Consequently for a given $f(x)$ the derivative $\varphi(x)$ can be defined in dozens of ways, even if we confine our attention to the simplest types of differential connections and consider only differentiation with respect to the independent variable. But if we introduce differentiation of one function with respect to another, i.e., consider the limits of a difference quotient of the form

$$\lim_{h \to 0} \frac{f_1(x+h) - f_1(x)}{f_2(x+h) - f_2(x)} = \varphi(x),$$

the number of "derivatives" increases greatly.

A second subquestion is connected with the word "exist" in the original question: does it mean the existence of the function $\varphi(x)$ at each point $x \in [a,b]$ for a given $f(x)$, $x \in [a,b]$, or shall we speak of the existence of $\varphi(x)$ even when we neglect some set of points in $[a,b]$? But one must then ask immediately which point sets can be neglected—finite, countable, of first category, of measure zero, of zero capacity, and so forth?

It is also appropriate at this point to take an interest in the admissible values of $\varphi(x)$, whether only finite values should be allowed, or whether one should speak of the existence of $\varphi(x)$ at a point under consideration even when it has an infinite value of a definite sign.

The number of such questions grows without limit if we consider not only functions of one variable assuming real values but also functions of several variables, set functions, functionals, operators, etc. None of these questions is idle—the majority of them have been posed and solved in analysis and the theory of functions, often with rather unclear formulations and methods of answering.

The most thoroughly studied case of a differential connection between $f(x)$ and $\varphi(x)$ is of course the connection defined by relation (1) with h varying continuously in both directions. Nearly all analysts from the seventeenth century on studied it to one degree or another; we shall use it to trace further the metamorphoses in the views of mathematicians, as indicated at the beginning of this section.

5.2 Ampère's theorem

The early analysts do not seem to have been interested in the question of the existence of a derivative for a hypothetical function $f(x)$. Their procedure was simpler—they calculated $f'(x)$, usually with success; exceptions occurred only at particular points of the domain of definition of $f(x)$. The practice of this type of computation throughout the seventeenth and eighteenth centuries grew into the conviction that in general every function has a derivative everywhere except at particular points. This conviction was strengthened by the practice of determining tangents for the curves under consideration and velocities for the motions being studied. In 1806 Ampère [1] attempted to give a theoretical justification for this conviction on a purely analytic basis. We shall pause to give the details of this attempt.

It is customary to begin a discussion of the question of the existence of nondifferentiable continuous functions with this work of Ampère. Following Hawkins [1, p. 43], we shall begin at an earlier time and allow ourselves more details, in order to avoid certain existing judgments of Ampère's paper that seem incorrect to us.

In 1772 Lagrange [1], starting from the expansion of $f(x+h)$ in a series of the form[1]

$$f(x+h) = f(x) + ph + p'h^2 + p''h^3 + \cdots \qquad (3)$$

---

[1] In [1] Lagrange was not yet using the symbol $f(x)$ to denote a function of one variable, and therefore he always wrote the expansion (3) without the left-hand side. Denoting the function under consideration by u and the increment of the independent variable x by ξ, he wrote (3)

where $p, p', p'', \ldots$ are functions of x alone, called p the derivative of the function $f(x)$ and denoted it by $f'(x)$. A similar meaning was attached to the derivatives of higher orders. From this he concluded that "the differential calculus, considered in full generality, consists of finding... the $p, p', p'', \ldots$ obtained from the function u, and the integral calculus consists of recovering u using the latter functions"[2] [1, p. 443].

In [1] Lagrange's goal was to give "the simplest possible" proof that the expansion (3) has the form

$$f(x+h) = f(x) + \frac{f'(x)}{1!}h + \frac{f''(x)}{2!}h^2 + \cdots, \qquad (4)$$

i.e., is the Taylor expansion.

In 1797, in his *Théorie des fonctions analytiques*, Lagrange again returned to the same questions, with the difference that where in [1] he had postulated the possibility of an expansion (3) with integer exponents, in [2] he took as an axiom the weaker assumption "verified by the expansion of many known functions" [2, p. 22] that fractional and negative powers of h may occur in (3). One of the essential additions in this regard consisted of his attempt to prove [2, pp. 22–24] that the increment of any function can be expanded in a series (3) containing only positive integer powers of h, neglecting a finite number of points at which the expansion breaks down and it is necessary to introduce fractional or negative powers. The other points of interest to us remained as before. He proceeded similarly in [3].

Thus the derivative $f'(x)$ of the function $f(x)$ is for Lagrange simply the coefficient of the term of first degree in the Taylor series expansion of its increment. Its existence everywhere except at a finite number of points is guaranteed by the initial axiom and the proof of the expansion (4); it requires no special proof. The individual points at which the expansion (4) breaks down are in addition the points at which the differential calculus itself becomes useless [2, p. 22].[3]

Ampère [1] took a higher view of things. To be specific, he introduced a definition of derivative independent of (4) and decided to prove the existence

---

simply in the form
$$u + p\xi + p'\xi^2 + p''\xi^3 + \cdots$$

[1, p. 442]. Later [2] he began to use the symbol $f(x)$, but continued to write (3) without the left-hand side, even though he used the notation $f(x+i)$.

[2] The text omitted from the quotation refers to functions of several variables, about which we shall not go into detail.

[3] This circle of questions has been discussed many times. Of the recent papers we mention, for example, the article of Fraser [1, pp. 39–44].

and uniqueness of the derivative for "every" function purely analytically, one of the first conscious attempts to state and prove a general existence theorem in analysis, not only in differential calculus. No matter how this attempt as a whole is judged, when we read in a paper of 1806 that, "this function [the derivative—F. M.]... is extremely important in mathematics and especially in its applications... and our first purpose is to prove that it exists" [1, p. 49], it is difficult not to be struck by the daring nature of the project; for there are not many general existence theorems in differential calculus even today, and in addition, at the beginning of the nineteenth century this calculus was the foundation of all infinitesimal analysis, so that such a proof would have strengthened the foundation, as Ampère explicitly emphasizes [1, p. 150].

It is difficult for the modern reader to judge the content of Ampère's main proposition and the methods of reasoning he used to prove it. The majority of authors who have written about this are for some reason inclined to believe that in this paper he was attempting to prove that a derivative exists everywhere except at a finite number of points for any continuous function or at least for every continuous function satisfying certain supplementary restrictions.[4] This opinion can hardly be considered correct. After all, the concept of a continuous function as it is known today, cannot be dated earlier than Lobachevskii and Dirichlet (cf. Sec. 8 of Essay II), or at most to Bolzano (1817) and Cauchy (1821). Thus it is clear a priori that Ampère could not have stated any such general considerations about such functions in 1806, and in fact he makes no such statements. He himself nowhere says explicitly that the functions he is considering belong to any general class of functions, and any judgment as to what he had in mind must be based on indirect reasoning.

The only mathematician to whom Ampère refers in [1] is Lagrange; he not only uses the terminology and notation of Lagrange, but also the methods of reasoning he applied. And when, for example, Ampère needed to carry out the delicate argument that a function assuming two different values assumes also a value intermediate between them, his only argument is the remark

---

[4] Cf., for example, Dini [4, p. 68], Pascal [1, p. 124], Pasch [3, p. 126]. Along with this one must not omit to mention the interesting opinion of F. Riesz on this score. Referring to the origin of Lebesgue's theorem on the existence of a derivative almost everywhere for a monotone continuous function, he wrote: "It dates from 1806, from the memoir of Ampère... in which that great scholar attempted to prove that "every" function is differentiable everywhere except at "individual and isolated" values of the variable. If we keep in mind the evolution of the idea of a function, we can conjecture—although the original text gives no positive indication of this—that the efforts of this great scholar could not have been directed outside the realm of functions formed from monotone pieces, i.e., outside the scope of the problem solved by Lebesgue" [12, p. 208]. Cf. also Grabiner [1, pp. 129–132].

that this proposition holds "by virtue of the reasoning used by Lagrange to prove the reality of irrational roots of algebraic equations that change sign when different numbers x and y are substituted in them" (Ampère [1, pp. 153–154]).

At the beginning of the nineteenth century Lagrange was perhaps *the* mathematical authority, and it therefore seems natural that Ampère remained basically within the limits of the mathematical conceptions of this great scholar.

But for Lagrange a function is analytic everywhere except at a finite number of isolated singular points and other kinds of functions simply do not exist. Therefore, taking account of what has been said above, it seems justified to conjecture that Ampère also, in talking about "any function," meant precisely these functions, and not arbitrary continuous functions. There is not the slightest mention of functions with finite discontinuities or continuous functions with a finite number of oscillations on a finite interval in Ampère's paper [1]. In addition he knows that a function may assume infinite values or vanish at individual points. The only thing that disturbs him is the question whether the function he is considering may become infinite or zero at every point, and this is the only thing he excludes from the outset [1, p. 148].[5]

If we accept this rather natural conjecture, both Ampère's basic proposition and his method of justifying it have an appearance different from the one usually presented: he is attempting to prove that every function that is analytic in the sense of Lagrange has a derivative everywhere except at individual isolated values of the variable, at which it may become infinite. But for such functions Ampère's proposition is completely correct, and the question rather arises as to why he is proving this.

The answer to this question again can be obtained by comparison with Lagrange. As we have said, the latter defined the derivative as the coefficient of the first-degree term of the unknown in the Taylor series expansion of the increment of the function, and its existence required no special proof once it was established that the function had such an expansion. But this definition of the concept of derivative was not the only one. From the moment differential calculus was conceived the derivative had been regarded in the final analysis as the function obtained using the differential relation (1), and it is in this form that it is constantly used in the calculation of derivatives. Even Lagrange used this notion of the derivative [2, p. 24], but only as an auxiliary device; in general he strove to avoid limit considerations, though not always successfully.[6]

---

[5] In particular it seems that for him $f(x) \equiv 0$ was not a function.

[6] Cf., for example, Lagrange [2, pp. 28–29].

Ampère, in contrast, introduced the concept of derivative itself, though not entirely clearly, using the limit (1). Indeed, in the introductory section of his paper he writes:

> It will necessarily follow from this proof that for $i = 0$ the expression
> $$\frac{f(x+i) - f(x)}{i}$$
> reduces to some function of x. This function, which obviously depends on x and which M. Lagrange has therefore called the derivative, is, as is known, extremely important in mathematics and especially in its applications in geometry and mechanics; following this illustrious mathematician, we shall denote it by $f'(x)$ ... [1, p. 148].

In these words we see an important deviation from the Lagrangian definition of the derivative, since the basis of this reasoning is the definition that was not only operational before Ampère but also became the basic one subsequently. To be sure there is no explicit concept of a limit, which is replaced by the rather vague (from our point of view) phrase "reduces for $i = 0$" but we shall not be too harsh on a paper of 1806—after all, Cauchy had not yet attempted the foundation of analysis on the concept of a limit. Still the idea of passing to a limit is also used more explicitly by Ampère, as can be seen, for example, from the following words:

> We first remark that for a real-valued function of x and i to vanish or become infinite for $i = 0$ this function must decrease or increase as i decreases in such a way that the function remains less than any given quantity for sufficiently small i in the first case and exceeds any given quantity when i is given a sufficiently small value in the second case" [1, p. 150].

Since the definition of the derivative introduced by Ampère was a new one compared with that of Lagrange, the proof of its existence, even for functions that are analytic in the sense of Lagrange was by no means a trivial problem.

We shall not continue our description of Ampère's memoir; we mention only that, relying on the theorem he had proved, he showed its usefulness in a variety of problems of analysis, geometry, and mechanics, and again gave a new proof of Taylor's theorem by studying the remainder.

Many of the authors who have written about this theorem, for example Dini and Pasch, have undoubtedly read this memoir of Ampère, but evidently under the influence of contemporary notions, have seen in it something that was not there. This attests yet again to the danger of modernization in the study of old works.

5.3 Doubts and refutations

Scholars seem to have received Ampère's theorem on the existence of a derivative everywhere except possibly at a finite number of points for an "arbitrary" function, as a natural bolstering of existing mathematical and scientific practice, which was confirmed by the computation of derivatives for specific functions, tangents for geometric curves, velocities for moving bodies, and the like. Here, for example, is what Poinsot wrote in 1815:

> One may even say that the ratio of two quantities of the same type depends neither on their nature, nor on their absolute values: by the very definition of ratio the quantity $(\Delta y : \Delta x)$ always has a limit; and it is precisely the existence of a curve and its tangent, whose presence undoubtedly shows the obviousness of the latter.[7]

It is difficult to say who deserves the dubious priority for carrying Ampère's proposition over to functions that are continuous in the modern sense. In any case in Raabe's book *Die Differential- und Integralrechnung mit Functionen einer Variabeln* [1], which appeared in 1839 this is done in all definiteness. After characterizing Ampère's theorem as the foundation for all differential calculus [1, p. v], and introducing the concept of a continuous function as was done by Cauchy in 1821 [1, p. 5], Raabe states and proves an existence theorem for the derivative of any continuous function [1, pp. 7–11], which he later used to establish other results. In the same year he applied this extended version of "Ampère's Theorem" to study the definite integral of a bounded function that was discontinuous or assumed infinite values at a finite number of points [2].

Raabe was not alone in doing this. On the contrary, variants of the same thing were done by Lacroix in 1810, Galois in 1831, Duhamel in 1847 and 1856, Lamarle in 1855, Freycinet in 1860, Bertrand in 1864, and Serre and Rubini in 1868.[8] Even in 1874 Weierstrass' student Königsberger was convinced of the correctness of this proposition under certain rather weak restrictive hypotheses,[9] and in their *Traité d'analyse* of 1878 Bertrand and

---

[7] Cited from the book of Brunschvicg [1, p. 337]. We call attention to the fact that the derivative is explicitly thought of as the limit of the ratio (1).

[8] For bibliographical references cf. Hawkins [1, pp. 44–45].

[9] Ibid., p. 46.

Garcet asserted it without any restrictions.[10] We shall not go into the details of the proofs of these authors, remarking only that it would be a mistake to characterize all these proofs as simply attempts with unsuitable means, since they involve a demonstrably false proposition. Rather refined devices were sometimes employed in their arguments, devices that were inspired precisely by the falsity of the proposition, but occasionally contain curious ideas. Thus Galois [1] began his proof with the difference quotient

$$\frac{F(x+h)-F(x)}{f(x+h)-f(x)}$$

and deduced from his argument "as a corollary that the quantity

$$\lim \frac{F(x+h)-Fx}{f(x+h)-fx}$$

is invariably some function of x for $h = 0$" [1, p. 9]. In other words, in 1831 Galois was talking about the derivative of one function with respect to another, a special case of which is the usual derivative with respect to the independent variable. But such a concept of the derivative was introduced into mathematics only in the twentieth century, except for some vague prefigurations of Cauchy (1841) and Peano (1887) of a slightly different order.[11] And if we regard this last equality as proved, the existence of a derivative of $F(x)$ in the sense of (1) can be immediately obtained from it by setting $f(x) = x$, as Galois remarks [1, p. 9].

But the time when mathematicians believed that every continuous function has a derivative everywhere except at a finite set of points was over. They began to recognize that continuity and differentiability of a function are not necessarily connected with each other. We have already mentioned briefly (p. 65) the function of Bolzano, the separation of continuity and differentiability by Lobachevskii, and the conviction expressed by Dirichlet that there exist continuous functions having a derivative at no point (p. 188). To what was said above we add the following.

The example of Riemann

$$f(x) = \sum_{n=1}^{\infty} \frac{(nx)}{n^2}$$

(1854, published in 1867) of a function that is discontinuous on a dense set of points, yet integrable, already contains implicitly an example of a function

---

[10] Cf. Brunschvicg, [1, p. 338].

[11] On these prefigurations cf., for example, Medvedev [1, 3].

that is nondifferentiable at the points of an everywhere dense set: the indefinite integral of $f(x)$ is such a function. In addition, around 1861 or earlier Riemann[12] exhibited the example of the function

$$f(x) = \sum_{n=1}^{\infty} \frac{\sin n^2 x}{n^2}$$

in reference to which Du Bois-Reymond asserted that it is also nondifferentiable on an everywhere dense set. Weierstrass gave a different opinion of this function and in regard to it he wrote in 1872:

> Until recently it was generally assumed that a single-valued continuous function of a real variable always has a first derivative whose value can become indeterminate or infinitely large only at individual points. Even in the letters of Gauss, Cauchy, and Dirichlet I have not succeeded in discovering indisputable evidence that these mathematicians, who subjected everything in their subject to rigorous criticism, expressed any different view. As I have learned from certain of Riemann's auditors, he was the first (around 1861 and possibly earlier) to express the thought that this conjecture is false, and, for example, that the function represented by the infinite series
>
> $$\sum_{n=1}^{\infty} \frac{\sin(n^2 x)}{n^2}$$
>
> refutes it. Unfortunately Riemann's proof was not published and does not seem to be contained in either his Nachlass or in oral transmission. This is the more regrettable since I have been able to verify many times that Riemann was attentive to his auditors. After Riemann's assertion had become widely known, the mathematicians studying this question (at least the majority of them) assumed that it suffices to prove the existence of a function that in every arbitrarily small interval has points at which it is nondifferentiable. The existence of functions of this type is very easy to prove, and I suppose therefore that Riemann had in mind functions that have no derivative for any value of their argument. The proof that a function of this type is a function representable by a trigonometric series still seems rather difficult to me... [5, pp. 71–72].

---

[12] Both Du Bois-Reymond in 1875 [3, p. 28] and Weierstrass in 1872 [5, p. 71, 1880 edition] communicated this with a reference to students of Riemann.

The difficulty of analyzing Riemann's second example is attested not only by Weierstrass' declining to carry it out, but also by the fact that until 1918 neither a proof nor a refutation of Riemann's assertion had appeared. Only Hardy in 1918, relying on certain delicate results of Diophantine analysis that he had previously obtained jointly with Littlewood, managed to show that this function of Riemann's does not have a finite derivative at any point of the form $\xi\pi$, where ξ is irrational or is a rational number of the form $\frac{2m}{4n+1}$ or $\frac{2m+1}{2(2m+1)}$, where m and n are integers [4, pp. 322–323], after which he generalized Riemann's example slightly [4, pp. 323–325]. This result of Hardy's was extended somewhat by Gerver [1] in 1970, who showed that it fails to have a finite derivative at the points $\xi\pi$, where ξ is a rational number of the form $\frac{2m+1}{2^n}$, where m and n are integers and $n \geq 1$. But at the same time he established that it does have a derivative (equal to $-1/2$) at the points $\xi\pi$, where ξ is a rational number with odd denominator and numerator, so that the Riemann function is differentiable on an infinite set of points.

The question of the differentiability of this function at the remaining rational numbers remained unclear. This question was solved by Gerver in a subsequent paper [2], where it is shown that there are no points of differentiability of the Riemann function except those already exhibited.

Until 1870 there seems to have been no published example of a continuous function having no derivative on an infinite set of points, not counting the first implicit example of Riemann, and Hoüel, in reviewing a memoir of Hankel [2] that appeared in that year and in which such functions were exhibited, commented that "there is no mathematician today who would believe in the existence of continuous functions without derivatives" [1, p. 123] and merely expresses the hope that Hankel's work will promote a change in this opinion. It is curious in this connection that the absence of a derivative at only rational points is identified with the absence of a derivative in general— a fact contained in the words of Weierstrass quoted above, and historically explicable, since the theory of point sets did not yet exist, but interesting when it is compared with modern notions, where, for example, the absence of a derivative on a set of points of measure zero by no means prevents us from talking about the derivative.

In the 1870's a crushing blow was delivered to the faith of mathematicians that a continuous function necessarily has a derivative, though perhaps not everywhere, and Hankel played a significant role in this. In 1870 he proposed the method of condensation of singularities [2, pp. 61–65], which consists of constructing a function using an absolutely convergent series, each term of which has a singular point; the function so obtained turns out to have these singularities at all rational points. It was by this method that he obtained the first examples of continuous functions having no derivative on the everywhere dense set of rational points.

One such example was the function

$$f(x) = \sum_{n=1}^{\infty} \frac{1}{n^s} \sin \pi n x \sin \left(\frac{1}{\sin \pi n x} \right),$$

where n is a natural number and $s > 1$.

In 1873 Schwarz [1] gave a different construction of a continuous function having no derivative on an everywhere dense set of points. To be specific, he started with the function

$$\varphi(x) = [x] - \sqrt{x - [x]},$$

where $x > 0$ and $[x]$ denotes the largest integer contained in x and considered the series

$$f(x) = \sum_{n=1}^{\infty} \frac{\varphi(2^n x)}{4^n}.$$

The function $f(x)$ so obtained turns out to be continuous and monotonically increasing, but fails to have a finite derivative at an infinite number of points on every interval. It is again curious that while we now say that this function is differentiable (almost everywhere), Schwarz regarded it as an example of a nondifferentiable function.

Weierstrass penetrated more deeply into this problem. It is not known exactly when he succeeded in constructing his famous example.[13] Weierstrass sent a communication about the continuous function

$$f(x) = \sum_{n=1}^{\infty} a^n \cos(b^n \pi x),$$

to the Berlin Academy of Sciences on 18 June 1872, but this example was published only in 1875 by Du Bois-Reymond [3, pp. 29–31]. Here b is an odd integer larger than 1, $0 < a < 1$, and $ab > 1 + \frac{3}{2}\pi$.

---

[13] Jourdain [1, p. 109] claimed that Weierstrass communicated it in lectures of 1861, referring to the paper of Schwarz [1, p. 269] for this. However in the place referred to Schwarz says only that in his 1861 lectures at the University of Berlin Weierstrass had spoken of the incorrectness of all investigations in which the existence of a derivative for every continuous function was proved; Weierstrass' example is not mentioned. The interpretation of differentiability by Schwarz just noted rather attests that when he wrote of this he did not know of Weierstrass' example. Cf. also Dugac [1, pp. 92–94].

Darboux had arrived at the same idea independently of Weierstrass.[14] He analyzed and generalized the examples of Hankel and Schwarz, and also constructed [1, pp. 107–108] the function

$$f(x) = \sum_{n=1}^{\infty} \frac{\sin[(n+1)!x]}{n!},$$

which does not have a derivative at any x.

The functions given as examples by Riemann, Hankel, Schwarz, Weierstrass, and Darboux evoked numerous investigations. These investigations were carried out in various directions: the examples themselves were studied, as has already been stated in part; new individual examples of functions nondifferentiable everywhere or on various infinite sets of points were constructed; a differentiation different from the one defined by relation (1) was considered, etc. The reason for the interest in papers of this type was that they damaged more and more the traditional notion of the connection between continuity and differentiability of functions; that the operation of differentiation, which had previously been considered comparatively simple and realizable, emerged as a different operation and was turning out to be frequently difficult to execute in practice, as could be seen from just the case of the Riemann function, whose study required delicate techniques of the theory of functions of a complex variable and Diophantine analysis; and finally that in the course of studying such functions new auxiliary techniques of analysis were created. As an example of papers of this last type, one can mention the article of Cahen [1], in which an interesting technique of representing real numbers generalizing the dyadic representation, is proposed.

5.4 Classes of nondifferentiable functions

In the development of the theory of differentiation in the nineteenth century the greatest contribution was undoubtedly due to the Italian mathematician Dini, whom we have mentioned many times in various connections. Besides the 1877 article [1] already mentioned, he published two other articles in the same year [2, 3], and a year later his famous *Fondamenti* [4] was published, in which his investigations on this problem were systematized. We shall digress slightly from the topic of this section in order to point out a few of his results.

---

[14] He had communicated his results at a meeting of the French Mathematical Society on 19 March 1873 and 28 January 1874, i.e., before the publication of Du Bois-Reymond. On the priority dispute between Weierstrass and Darboux cf. Dugac [1, pp. 93–94].

5.4 Classes of nondifferentiable functions

The following theorem of Dini deserves to be mentioned first:

If $f(x)$ is a function that is finite and continuous on a whole interval (including α and β) and for all points x between α and β (except possibly α and β) it has a finite well-defined derivative, or possibly an infinite one of definite sign, then in any portion of this interval this derivative:

1) cannot have infinite values at every point;

2) cannot be identically zero unless the function is constant on that portion of the interval;

3) cannot have only 0 and ∞ as values [4, p. 69].

It was not until 1908 that Cahen [1, pp. 212–213], without referring to Dini, restated this proposition as the assertion that in general there do not exist continuous functions whose derivatives are everywhere infinite, everywhere zero (except $f(x) = $ const), or assume only zero and infinity as values, and gave it a slightly different proof. In 1912 Luzin [1, pp. 5–11] succeeded in generalizing the main part of this theorem to assert that there does not exist a continuous function $F(x)$ such that $\left|\dfrac{dF}{dx}\right| = +\infty$ on a set of positive measure.

Dini next proved [4, pp. 71–72] that under the same assumptions on $f(x)$ this function always has a nonzero finite derivative on an everywhere dense set—a fact that deserves to be mentioned because the theory of point sets had only begun to be created at the time.

It is also of interest that Dini came near to Lebesgue's theorem on the derivative of a continuous monotonic function. Indeed, in one place we read

> ...functions $f(x)$ that are always increasing or always decreasing on a given interval or assume only a finite number of maxima and minima on that interval, when increased by a linear function, must necessarily have a definite (finite or infinite) derivative from both the right and left at each point of the interval, and in any portion of the interval there must exist other finite intervals at which both the right- and left-hand derivatives, besides having a definite value, must be finite. However, these derivatives, if they are not everywhere continuous, may be different from the two sides of some points in every interval, although in each portion of this interval there must exist an infinite set of points at which the right- and left-hand derivatives, besides having a definite finite value, are also equal, so that at these points the ordinary derivative exists and is finite [4, p. 11].[15]

---

[15] This formulation had been published earlier by Dini in [2, p. 11].

At the level of knowledge of the theory of point sets possessed by Dini in 1877, one could hardly have said any more about the derivatives of monotonic functions.

Much more could be said about the differential properties of continuous functions studied by Dini (it suffices to say that more than a hundred pages of [4] are devoted to these properties), but we shall confine ourselves to what has already been said and turn to the question of immediate interest to us.

In the examples of continuous nondifferentiable functions just given, the main topic was individual functions, despite the fact that parameters were present. Dini took a more general point of view: in [1] he formulated, and in the following year [4, pp. 148–153] proved, for example, the following rather general existence theorem for continuous functions that do not have a finite or infinite derivative at any point:

Suppose that functions $f_n(x)$ are defined on $0 \leq x \leq 1$ satisfying the following conditions:

1) all the functions $f_n(x)$ are continuous and have bounded derivatives;

2) the series

$$\sum_{n=1}^{\infty} f_n(x) \qquad (5)$$

converges on $[0,1]$ to a continuous function $f(x)$;

3) each of the $f_n(x)$ has a finite number of extrema, and the number of these extrema increases with n in such a way that for any $\varepsilon > 0$ there exists n_0 such that for $n > n_0$ the distance between successive extrema is less than ε;

4) if δ_n is the greatest distance between two successive extrema and D_n is the difference of two successive extreme values that is largest in absolute value, then

$$\lim_{n \to \infty} \frac{\delta_n}{D_n} = 0;$$

5) if h_n denotes for each x the two increments (one positive, one negative) for which $x + h_n$ gives the first right (resp. left) extremum for which

$$|f_n(x + h_n) - f_n(x)| \geq \frac{1}{2} D_n,$$

then it is possible to define positive numbers r_n such that

$$|R_n(x + h_n) - R_n(x)| \leq 2r_n$$

for all $x \in [0,1]$ *and the* h_n *corresponding to each* x*, where* $R_n(x)$ *is the remainder of the series* (5);

6) *if* c_n *is a sequence of positive numbers such that* $|u'_n(x)| < c_n$ *for all* $x \in [0,1]$*, then from some index on*

$$\frac{4\delta_n}{D_n}\sum_{\nu=1}^{n} c_\nu + \frac{4r_n}{D_n} \le \theta, \quad 0 \le \theta < 1;$$

7) *the sign of the difference* $f_n(x+h_n) - f_n(x)$ *is independent of* h_n *from some* n_0 *on for all* x.[16]

Then the function $f(x)$ *defined by the series* (5) *does not have a finite derivative at any point* $x \in [0,1]$*. It may have an infinite derivative on an infinite set of points.*

Dini then showed [4, pp. 169–171] that under certain additional assumptions such a function $f(x)$ will not even have an infinite derivative at any point.

The class of functions defined by this theorem of Dini's contains an infinite number of functions, as Dini specifically emphasized [1, p. 6]; in particular it contains the Weierstrass function (Dini [4, p. 162]).

After Dini a rather general method of constructing a class of functions having no derivative was proposed by Darboux [2]. His method is clearer. He studied functions $\varphi(x)$ defined by a series of the form

$$\varphi(x) = \sum_{n=1}^{\infty} \frac{f(a_n b_n x)}{a_n}, \tag{6}$$

where a_n and b_n are sequences of real numbers and $f(x)$ is a bounded continuous function with bounded second derivative. If the sequences $\{a_n\}$ and $\{b_n\}$ are chosen so that for fixed k

$$\lim_{n\to\infty} \frac{a_n}{a_{n+1}} = 0, \quad \lim_{n\to\infty} \frac{a_1 b_1^2 + a_2 b_2^2 + \cdots + a_{n-k} b_{n-k}^2}{a_n} = 0,$$

then the series (6) converges everywhere to a continuous function $\varphi(x)$.

Under further restrictions on the choice of $\{a_n\}$, $\{b_n\}$, k, and $f(x)$ one can obtain continuous functions having a derivative at no point. Thus, if $b_n = 1$ and $k = 1$, it suffices to impose on the numbers a_n the condition

$$\lim_{n\to\infty} \frac{a_1 + a_2 + \cdots + a_{n-1}}{a_n} = 0,$$

---

[16] Following Knopp [1, pp. 23–25], we have modernized Dini's statement of this result somewhat.

which is satisfied, for example, by the numbers $a_n = n!$, in order to exhibit an infinite set of functions $f(x)$ for which the $\varphi(x)$ defined by Eq. (6) is such a function. If, for example, under these last assumptions $f(x) = \cos x$, then $\varphi(x)$ is nowhere differentiable; under the choice $b_n = n+1$, $k = 3$ and $f(x) = \sin x$ the function $\varphi(x)$ of Darboux' previous memoir is obtained.

After the memoirs of Dini and Darboux an entire industry, if one may so express it, was created for the production of both individual nondifferentiable functions and classes of such functions.

The interest in these investigations was stimulated not only by the factors mentioned at the end of the preceding section, but also by the fact that the discovery of infinite classes of continuous functions that are nondifferentiable in the ordinary sense suggested looking for conditions for differentiability of a function on the one hand and led to the idea of generalizing the idea of differentiation itself on the other hand. In 1877 Dini wrote frankly that "it is necessary to conduct general studies to make clear the restrictions that must be introduced into the concept of a function in order for the differential calculus to be applicable to it, or to find more general methods of calculation (di calcolo) applicable to any continuous function" [2, p. 8]. Here he formulates the concept of the derivates of a continuous function, to whose study he devoted many pages of his monograph [4].

We do not intend to study these questions, but instead we shall return to the final step (in a certain sense) on the path of mathematicians in the recognition of the place occupied by differentiable functions in the set of all continuous functions.

5.5 The relative "smallness" of the set of differentiable functions

We begin by giving the definition of sets of first and second category. Since we shall later have to deal with nonnumerical sets, we give an abstract definition, relying on the book of Kuratowski [2].

A set M is called a *topological space* if to each subset $X \subset M$ there corresponds a subset $\overline{X} \subset M$ such that the following axioms hold:

1) $\overline{X \cup Y} = \overline{X} \cup \overline{Y}$,
2) $X \subset \overline{X}$,
3) $\overline{\emptyset} = \emptyset$, where $\emptyset$ is the empty set.
4) $\overline{\overline{X}} = \overline{X}$ [2, Vol. I, pp. 20].[17]

---

[17] We call attention to the fact that Kuratowski calls the correspondence $X \to \overline{X}$ a function, meaning a function in the sense of Carathéodory (cf. p. 80 above).

The set X is said to be *everywhere dense* in the topological space M if $\overline{X} = M$. A set X is said to be *nowhere dense* if $\overline{M \setminus \overline{X}} = M$ [2, Vol. I, p. 36].

A set is said to be *of first category* if it is the union of a countable family of nowhere-dense sets [2, Vol. I, p. 48]. In many problems of the theory of functions and topology sets of first category play a role analogous to that of sets of measure zero in measure theory—they can be neglected.

As for sets of second category, they are defined in different ways, not all of them equivalent. Baire himself defined them as sets that are not of first category [3, p. 1623; 5, p. 65; 7, p. 83]. Hausdorff [1, p. 142] and Saks [2, p. 41] also introduced them this way. In 1900 Schönflies [1, pp. 108–109] defined them as sets that are the complements of sets of first category, and Hobson and other mathematicians followed him in doing so.[18] In the Russian mathematical literature the second definition seems to have been more common in the earlier period (cf., for example, Aleksandrov [4, p. 167] and Bari and Men'shov [1, p. 392]), while the first came to be more often applied later on (cf., for example, Kantorovich and Akilov [1, p. 26]). At present sets of second category are defined according to Baire's definition (cf. Oxtoby [1, p. 2]), and the complement of a set of first category is called a *residual set* (loc. cit., p. 41).

The question of the relation between the set of all continuous functions and the subset of it consisting of differentiable functions was posed in 1929 by Steinhaus. In 1931 it was solved using slightly differing methods by S. Mazurkiewicz and Banach, and their result was supplemented a year later by Saks. Before discussing this we shall make another digression.

Existence theorems in set theory and the theory of functions can be proved by various methods. One method, for example, is based on the concept of cardinality: to prove that a set E contains an element that does not possess a certain given property one proves that E has more elements in the sense of cardinality than those possessing the given property. Thus as early as 1874 Cantor gave the following proof that there exist transcendental numbers: it follows from the fact that the set of real numbers is uncountable and the set of algebraic numbers is countable that there exist transcendental numbers. Moreover the number of algebraic numbers is negligibly small (with respect to cardinality) in comparison with the set of transcendental numbers.

Another method of this type is the method of measure, based on comparing sets according to their measure. If, for example, the set A of points of the interval $(0, 1)$ possessing the property P is such that one can prove the inequality $mA < 1$, then there are demonstrably points on $(0, 1)$ not possessing property P; such points even form an uncountable set.

---

[18] For details cf. Rosenthal [1, p. 86, footnote 99].

The third method, which contains the two preceding methods as special cases, is the method of category, i.e., a method of proving existence theorems based on Baire's Theorem, according to which a complete metric space is a set of second category upon itself.[19] Introducing the principle of comparison of sets in accordance with their category, one can reason as follows: suppose it is required to prove that there exists an element x in a complete metric space M not possessing the property P; if we succeed in proving that the set R of elements $x \in M$ possessing property P can be represented as a union $R = \bigcup_{i=1}^{\infty} R_i$, where each R_i is nowhere dense in M, then the set R, being a set of first category, cannot coincide with M, i.e., there exist elements in M not possessing the property P.

The method of category originated in the memoir of Baire [3]. Baire used this method, in particular, in proving his theorem on functions of first class.[20] Lebesgue applied this method many times.[21] But it seems to have been only after the appearance in 1917 of Lebesgue's memoir "Sur certaines démonstrations d'existence" [17] in which all three of these methods were described generally and compared with one another that the last of them began to be applied more often, especially starting with the works of the Polish school of set theory and the theory of functions.

It was by using arguments of exactly this type that S. Mazurkiewicz [2] and Banach [1] proved the following theorem.

Let C be the space of continuous functions x having period 1, endowed with the usual norm $\|x\| = \max |x(t)|$, $0 \leq t \leq 1$. Suppose also that N is the set of functions in C having no finite right-hand derivative at any point $t \in [0, 1]$. Then N is a set of second category in C, and its complement is a set of first category.

Thus the set of functions having a finite one-sided derivative at even one point $t \in [0, 1]$ is negligibly small in the sense of Baire category in comparison with the set of all continuous functions. This holds *a fortiori* for the set of functions with a finite ordinary derivative.

Before taking up the work of Saks [1], we make one more digression. Neither the individual continuous nondifferentiable functions nor the classes of such functions studied in the nineteenth century and the first two decades of the twentieth century provided an example of a continuous function so singular as to have neither a finite nor infinite one-sided derivative at any point. The Weierstrass function given above, for example, has a one-sided

---

[19] Cf. Saks, [2, pp. 54–55].
[20] Cf., for example, Baire [7, pp. 87–88].
[21] Cf., for example, Lebesgue [10, pp. 159–160, 184–186].

derivative on an everywhere-dense set. The first example of such a function was not found until 1922 when Besicovitch [1, pp. 548–556] constructed one (and published it in 1924).

It is precisely in connection with this function that Banach and Steinhaus posed the following question: Can the result of S. Mazurkiewicz and Banach be extended to functions of Besicovitch type using the method of category, i.e., can it be shown that the complement of the set of all continuous functions having neither a finite nor infinite derivative at any point is of first category?

Saks gave a negative answer to this question in 1932. He established [1, pp. 215–217] that the set of continuous functions on $[0, 1]$ for which there exists either a finite or positive-infinite right-hand derivative on a set of cardinality c is of second category in the space of all continuous functions.

Consequently the class of functions having a one-sided derivative at at least one point is significantly larger in the sense of category than the class of functions having an ordinary derivative at at least one point. Accordingly the class of functions having neither a finite nor an infinite one-sided derivative at each point of its domain of definition is smaller in the same sense than the class of functions having no two-sided derivative. "This perhaps explains the difficulty of finding the first example of a function having neither a finite nor infinite one-sided derivative at every point" (Saks [1, p. 212]). In addition Saks' result points up the essential difference between two-sided and one-sided differentiation.[22]

We shall say a few more words about the paper of Orlicz [1].

The proofs of S. Mazurkiewicz, Banach, and Saks are nonconstructive in the sense that the reasoning establishes the mere fact that the function under consideration exists without giving any method of representing it analytically. Their methods can lead only indirectly to specific examples of nowhere differentiable functions. In addition many investigators beginning with Dini and Darboux directed their efforts toward obtaining such nondifferentiable functions using series or sequences of functions, and in this situation the tendency to make such a method of definition as general as possible so as to encompass as many nondifferentiable functions as possible can be clearly traced. Orlicz also set himself such a goal [1].

He found some rather general conditions under which definite uniformly convergent series of continuous functions have nondifferentiable continuous functions as sums. But the greater generality of the class of such analytically representable functions that Orlicz attained in comparison with his predecessors was achieved at the expense of giving a noneffective definition of the coef-

---

[22] The results of Banach, Besicovitch, and Saks were expounded by Oxtoby [1, pp. 45–46].

ficients of the functional series he was studying, using the method of category. He described this method [1, p. 46] as "in some sense intermediate" between the "effective" methods of defining nondifferentiable functions as series and the "noneffective" method of S. Mazurkiewicz and Banach. The method of category was later applied in many questions of the theory of functions, including differentiability questions. We shall stop at this point, however, and give only a summary of what has been said above.

After mathematicians had discovered the operation of differentiation in its simplest form, circumstances arranged themselves favorably for them in the sense that they knew how to differentiate every function they were presented with. This created the impression that in general any function could be differentiated, not only once, but as many times as desired; exceptions arose only at individual points. The clearest exponent of this point of view at the turn of the nineteenth century was Lagrange, and Ampère attempted a theoretical justification of it. In broad outline this was correct, since mathematicians simply did not know functions of any other kind.

In the nineteenth century the main object of study for analysts was formed by the continuous functions. This class of functions is much larger than the class of functions studied earlier, and there was no reason to expect *a priori* that the operation of differentiation, which was adapted for the study of analytic functions, would turn out to be suitable for the study of continuous functions as well. However, since no new differential operation had been discovered, mathematicians applied the old one to functions of a more general nature. In this they were misled by the fact that the concept of continuous function in general includes the functions studied earlier, and differentiations, which it seemed could be performed on an arbitrary function, could actually be realized on functions of the earlier type. This led to the conviction that every continuous function is also differentiable. This conviction was strengthened by the "proofs" of authoritative scholars and on the whole was predominant until the 1870's. It seemed the more plausible because the representatives of mathematical science were satisfied by the existing differential apparatus: the motions they studied were described by continuous functions that were differentiable the necessary number of times.

In the 1870's a crushing blow was delivered to the belief that an arbitrary continuous function was differentiable: not only examples of individual continuous functions having a derivative at no point but whole (infinite) classes of such functions were constructed. The study of the new mathematical objects encountered the resistance of mathematicians of a traditional bent of mind. "I turn away in horror and disgust from this growing plague of nondifferentiable functions," wrote Hermite to Stieltjes (Saks [2, p. iv]). Even at the beginning of the twentieth century Boussinesq was not alone in expressing the opinion

5.5 The relative "smallness" of the set of differentiable functions

that "all the interest of a function depends on its having a derivative" (Denjoy, Felix, Montel [1, p. 15]), meaning an ordinary derivative. In Saks' words, "researches dealing with non-analytic functions and with functions violating laws which one hoped were universal were regarded almost as the propagation of anarchy and chaos where past generations had sought order and harmony" [2, p. iv].

By the beginning of the twentieth century the representatives of the traditional view had begun to yield their positions, though not completely. It was difficult to combat obvious facts and a curious temporary exit was found. Functions having singularities, such as very discontinuous functions or continuous functions without a derivative, came to be regarded as annoying exceptions among the "good" functions, as certain pathological phenomena in the basically healthy body of analysis, and it was even generously decided to study these diseased tumors, since "just as human pathology has its laws, so one may assume the presence of certain governing regularities in the singular properties of functions; in both the one and the other scientific understanding will have been attained when we succeed in passing from individual singular phenomena to the knowledge of the laws governing them" (Schönflies, [2, p. 111]).

It turned out in the 1930's that this picture did not correspond to the actual state of affairs. The class of continuous nowhere differentiable functions turned out to be immeasurably richer than the class of differentiable functions and it was rather functions of the latter type that were "pathological." A curious situation arose, when it turned out that the continuous functions that had been studied by mathematicians for centuries, those that were used to describe the phenomena of the external world, belong to a negligibly small class of continuous functions. It resembles a great deal the situation with irrational numbers: the discovery of incommensurables in Ancient Greece corresponds to the discovery of particular nondifferentiable functions; the discovery, also in Greece, of infinite classes of irrational numbers corresponds to the discovery of infinite classes of nondifferentiable functions; finally, Cantor's proof that the set of irrational numbers is uncountable, while the set of rational numbers is only countable, corresponds to the discovery by S. Mazurkiewicz' and Banach that the set of nondifferentiable functions is a set of second category in the space of all continuous functions, while the differentiable functions form only a set of first category in this space.

It should be mentioned that there are even some physical phenomena that it has turned out to be more convenient to describe in the language of nondifferentiable functions. As early as the beginning of the twentieth century Perrin remarked that the extremely irregular trajectories of particles in Brownian motion resemble nondifferentiable functions. This remark was taken up by Wiener and developed into a new theory of Brownian motion. "Under this theory I was able confirm the conjecture of Perrin and to show that, except for a set of cases of total probability 0, all the Brownian motions were continuous non-differentiable curves" [1, p. 39]. This is not the only type of motion in which this situation holds.

Bibliography

Abbreviations used in the bibliography

AM – *Annals of Mathematics*
BAMS – *Bulletin of the American Mathematical Society*
DAN – *Doklady Akademii Nauk* (English translation by the American Mathematical Society: *Soviet Math: Doklady*)
JLMS – *Journal of the London Mathematical Society*
IMI – *Istoriko-Matematicheskie Issledovaniya*
CR – *Comptes Rendus Hebdomadaires des Séances de l'Académie des Sciences (de Paris)*
MA – *Mathematische Annalen*
Matem. Sb. – *Matematicheskii Sbornik*
PAMS – *Proceedings of the American Mathematical Society*
PLMS – *Proceedings of the London Mathematical Society*
TAMS – *Transactions of the American Mathematical Society*
FM – *Fundamenta Mathematicæ*
UMN – *Uspekhi Matematicheskikh Nauk* (English translation by the American Mathematical Society: *Russian Mathematical Surveys*)

Abel, N. H.

1. 'Recherches sur la série $1 + \frac{m}{1}x + \frac{m(m-1)}{1\cdot 2}x^2 + \frac{m(m-1)(m-2)}{1\cdot 2\cdot 3}x^3 + \cdots$', (1826) *Œuvres Complètes*, T. 1, Christiania (Oslo), 1881, 219–250.

Agnesi, M. G.

1. *Traités élémentaires de calcul différentiel et de calcul intégral* (1748), Paris, 1775.

Agostini, A.

1. 'Il concetto d'integrale definito in Pietro Mengoli', *Period. Matem.*, ser. 4, **5**, 1925, 137–146.

Aleksandrov, P. S.

1. 'Über die Äquivalenz des Perronschen und Denjoyschen Integralbegriffes', *Math. Z.*, **20**, 1924, 213–222.
2. 'L'intégration au sens de M. Denjoy considérée comme recherche des fonctions primitives', *Matem. Sb.*, **31**, 1924, 465–476.
3. 'On the so-called quasiuniform convergence' [Russian], *UMN*, **3**, No. 1(23), 1948, 213–215.
4. *Einführung in die Mengenlehre und die Theorie der reellen Funktionen*, VEB Deutscher Verlag der Wiss., Berlin, 1967.

Ampère, A. M.

1. 'Recherches sur quelques points de la théorie des fonctions dérivées qui conduisent à une nouvelle démonstration de la série de Taylor, et à l'expression finie des

termes qu'on néglige lorsqu'on arrête cette série à un terme quelconque', *J. de l'Ec. Polyt.*, **6**, No. 13, 1806, 148–181.

Antropova, V. A.
1. 'On a geometric method of Newton's *Principia*' [Russian], *IMI*, **17**, 1966, 205–228.

Arzelà, C.
1. 'Un teorema intormo alla serie di funzioni', *Atti della R. Accad. dei Lincei*, Rend. Cl. delle sci. fis., matem. e nat., ser. 4, **1**, 1885, 262–267.
2. 'Sulla integrabilità di una serie di funzioni', Ibid., 321–326.
3. 'Sulla integrazione per serie', Ibid., 532–537, 566–569.
4. 'Sulla integrazione per serie', Ibid., ser. 5, **6**, 1897, 2nd semester, 290–292.
5. 'Sulle serie di funzioni', *Mem. della R. Accad. delle Sci. dell'Ist. di Bologna*, ser. 5, **8**, 1899–1900, 130–186, 701–744.

Ascoli, G.
1. 'Sul concetto di integrale definito', *Atti della R. Accad. dei Lincei*, Rend., Cl. delle sci. fis., matem. e nat., ser. 2, **2**, 1875, 862–872.

Averbukh, V. I. and Smolyanov, O. G.
1. 'Theory of differentiation in topological linear spaces' [Russian], *UMN*, **22**, No. 6(138), 1967, 201–260.
2. 'Different definitions of the derivative in topological linear spaces' [Russian], *UMN*, **23**, No. 4(142), 1968, 67–116.

Baire, R.
1. 'Sur la théorie générale des fonctions de variables réelles', *CR*, **125**, 1897, 691–694.
2. 'Sur les fonctions discontinues développables en séries de fonctions continues', Ibid., **126**, 1898, 884–887.
3. 'Sur les fonctions discontinues qui se rattachent aux fonctions continues', Ibid., 1621–1623.
4. 'Sur la théorie des fonctions discontinues', Ibid., **129**, 1899, 1010–1013.
5. 'Sur les fonctions de variables réelles', *Ann. Matem. Pura ed Appl.*, ser. 3, **3**, 1899, 1–123.
6. 'Sur les séries à termes continus et tous de même signe', *Bull. Soc. Math. France*, **32**, 1904, 125–128.
7. *Leçons sur les fonctions discontinues*, Paris, Gauthier-Villars, 1930.

Banach, S.
1. 'Über die Baire'sche Kategorie gewisser Funktionenmengen', *Stud. Math.*, **3**, 1931, 174–149.
2. *A course of Functional Analysis (Linear Operations)* [Ukrainian] (1932), Ryadanska Shkola, Kiev, 1948.

Bari, N. K. and Men'shov, D. E.
1. Commentary to the book of N. N. Luzin: *The Integral and the Trigonometric Series*, Gostekhizdat, Moscow-Leningrad, 1951, 389–537.

Bari, N. K., Lyapunov, A. A., Men'shov, D. E, and Tolstov, G. P.
1. 'The metric theory of functions of a real variable' [Russian], in: *Mathematics in the USSR after 30 years, 1917-1947*, Gostekhizdat, 1948, 256-318.

Baron, M. E.
1. *The origins of the infinitesimal calculus*, Oxford, London, Edinburgh, New York, Toronto, Sydney, Paris, Braunschweig, 1969.

Bashmakova, I. G.
1. 'Les méthodes différentielles d'Archimède', *Arch. Hist. Exact Sci.*, **2**, 1964, 87-107.

Bashmakova, I. G. and Yushkevich, A. P.
1. 'Leonhard Euler' [Russian], *IMI*, **7**, 1954, 451-512.

Bauer, H.
1. 'Der Perronische Integralbegriff und seine Beziehung zum Lebesgueschen', *Monatsh. Math. u. Phys.*, **26**, 1915, 153-198.

Bezikovich, A. S. (Besicovitch, A. S.)
1. 'An investigation of continuous functions in connection with the question of their differentiability' [Russian], *Matem. Sb.*, **31**, 1924, 529-556.

Bendixson, I.
1. 'Sur la convergence uniform des séries', *Öfversigt af Kongl. Vetenskaps-Akademiens Förhandlingar*, Stockholm, 1897, 605-622.

Bernkopf, M.
1. 'The development of function spaces with particular reference to their origins in integral equation theory', *Arch. Hist. Exact Sci.*, **3**, No. 1, 1966, 1-96.

Bieberbach, L.
1. 'Über einen Osgoodschen Satz aus der Integralrechnung,' *Math. Z.*, **2**, 1918, 155-157, 474.

Birnbaum, Z. W. and Orlicz, W.
1. 'Über die Approximation im Mittel', *Stud. Math.*, **2**, 1930, 197-206.
2. 'Über die Verallgemeinerung des Begriffes der zueinander Konjugierten Potenzen', Ibid., **3**, 1931, 1-67.

Biryukov, B. V.
1. 'On the works of Frege on the philosophical questions of mathematics' [Russian], in: *Philosophical Problems of Science*, V. II, (1957), Moscow University Press, 134-177.
2. *The Collapse of the Metaphysical Conception of Universality of Extension in Logic* [Russian], Vysshaya Shkola, Moscow, 1963.

Bochner, S.
1. 'The signification of some basic mathematical conceptions for physics', *Isis*, **54**, No. 2, 1963, 179-205.

Bolzano, B.

1. 'Rein analytischer Beweis des Lehrsatzes, daß zwischen je zwei Werten, die entgegengesetztes Resultat gewahren, wenigstens eine reele Wurzel der Gleichung liege', *Prag. Abh. der Böhm. Gesell.*, **V**, 1818.

Borel, E.

1. 'Sur quelques points de la théorie des fonctions', *Ann. Sci. Ec. Norm. Sup.*, sér. 3, **12**, 1895, 9–55.
2. *Leçons sur la théorie des fonctions*, Paris, 1898.
3. 'Un théorème sur les ensembles mesurables', *CR*, **137**, 1903, 966–969.
4. *Leçons sur les fonctions de variables réelles et les développements en séries de polynomes*, Paris, 1905.
5. 'Sur la définition de l'intégrale définie', *CR*, **150**, 1910, 375–377.
6. 'Sur les théorèmes fondamentaux de la théorie des fonctions de variables réelles', Ibid., **154**, 1912, 413–415.
7. 'Le calcul des intégrales définies', *J. math. pures et appl.*, sér. 6, **8**, 1912, 159–210.
8. 'La théorie de la mesure et la théorie de l'intégration', in: *Leçons sur la théorie des fonctions*, 2nd ed., Paris, 1914, 217–256.

Bortolotti, E.

1. 'La memoria "De infinitis hyperbolis" di Torricelli', *Archeion*, **6**, 1925, 49–58, 139–152.
2. *La storia della matematica nella Università di Bologna*, Bologna, 1947.

Bourbaki, N.

1. *Eléments de Mathématique*, XVII, 1ère partie, Livre I, Théorie des ensembles, Hermann, Paris, 1954.
2. *Eléments de Mathématique*, IX, 1ère partie, Livre IV, Fonctions d'une variable réele, Hermann, Paris, 1949.
3. *Eléments d'histoire des mathématiques*, Masson, Paris, 1984.

Boyer, C. B.

1. *The History of Calculus and its Conceptual Development*, New York, 1959.

Braun, Ben-Ami

1. 'An extension of a result by Talalyan on the representation of measurable functions by Schauder bases', *PAMS*, **34**, 1972, 440–446.

Brunschvicg, L.

1. *Les étapes de la philosophie mathématique*, Paris, 1912.

Bržečka, V. F.

1. 'On Bolzano's function' [Russian], *UMN*, **4**, No. 2(30), 1949, 15–21.

Burkill, J. C.

1. 'Functions of intervals', *PLMS*, ser. 2, **22**, 1924, 275–310.
2. 'The strong and weak convergence of functions of general type', Ibid., ser. 2, **28**, 1928, 493–500.

Cahen, E.
1. 'Sur une fonction continue sans dérivée', *Ann. Sci. de l'Ec. Norm. Sup.*, sér. 3, **25**, 1908, 199–219.

Cantor, G.
1. 'Beweis, daß eine für jeden rellen Wert von x durch eine trigonometrische Reihe gegebene Funktion $f(x)$ sich nur auf eine Weise in dieser Form darstellen läßt' (1870), *Ges. Abh.*, Berlin, 1932, 80–83.
2. 'Notiz zu dem Aufsatze:"Beweis, daß..."' (1871), Ibid., 84–86.
3. 'Über trigonometrische Reihen' (1871), Ibid., 87–91.
4. 'Ein Beitrag zur Mannigfaltigkeitslehre' (1878), Ibid., 119–133.
5. 'Über unendliche linear Punktmannigfaltigkeiten' (1878–1884), Ibid., 139–244.
6. 'Fernere Bemerkung über trigonometrische Reihen' (1880), Ibid., 104–106.
7. *Grundlagen einer allgemainen Mannigfaltigkeitslehre*, Leipzig, 1883.
8. 'Beiträge zur Begründung der transfiniten Mengenlehre' (1895–1897), *Ges. Abh.*, Berlin, 1932, 282–351.

Cantor, M.
1. *Vorlesungen über Geschichte der Mathematik*, Bd. III, Leipzig, 1901.

Carathéodory, C.
1. 'Vorlesungen über reellen Funktionen', Leipzig-Berlin, 1918.
2. 'Entwurf für eine Algebraisierung des Integralbegriffs' (1938), *Ges. Abh.*, Bd. IV, München, 1956, 302–342.
3. 'Über die Differentiation von Maßfunktionen' (1940), Ibid., 385–396.
4. 'Maß und Integral und ihre Algebraisierung', Basel-Stuttgart, 1956.

Carnot, L.
1. 'Réflexions sur la métaphysique du calcul infinitésimal', Duprat, 1797.

Cauchy, A. L.
1. *Analyse Algébrique*, 1ère partie du Cours d'analyse de l'Ecole royale Polytechnique, Imprimérie Royale, Paris, 1821.
2. *Résumé des leçons de calcul infinitésimal*, Imprimérie Royale, Paris, 1823.
3. 'Mémoire sur le rapport différentiel de deux grandeurs qui varient simultanement' (1841), *Œuvres Compl.*, sér. II, T. XII, Paris, 1916, 214–262.
4. 'Note sur les séries convergentes dont les divers membres sont des fonctions continues d'une variable réelle ou imaginaire, entre des limites données' (1853), *Œuvres Compl.*, sér. I, T. XII, Paris, 1900, 30–36.

Cavalieri, B.
1. *Exercitationes Geometrica Sex* (1647), in: *Geometria indivisibilibus continuorum nova quadam ratione promota*, 2nd ed., Bologna, 1653.

Church, A.
1. *Introduction to Mathematical Logic*, Princeton University Press, 1956.

Cooke, R. G.
1. *Infinite Matrices and Sequence Spaces*, Macmillan, London, 1949.

D'Alembert, J. L.

1. 'Sur un paradoxe géométrique', *Opuscules mathématiques*, IV, Paris, 1768, 62–65.

Daniell, P. J.

1. 'Differentiation with respect to a function of limited variation', *TAMS*, **19**, 1918, 353–362.
2. 'Stieltjes derivatives', *BAMS*, **26**, 1919/20, 444–448.
3. 'Derivatives of a general mass', *PLMS*, ser. 2, **26**, 1927, 95–118.
4. 'Stieltjes derivatives', Ibid., ser. 2, **30**, 1930, 187–198.

Dannemann, F.

1. *Grundriss einer Geschichte der Naturwissenschaften*, Bd. II, Leipzig, 1898. in the Sixteenth and Seventeenth Centuries, T. II, Macmillan, New York, 1935.

Darboux, G.

1. 'Mémoire sur les fonctions discontinues', *Ann. Sci. Ec. Norm. Sup.*, sér. 2, **4**, 57–112.
2. 'Addition au mémoire sur les fonctions discontinues', Ibid., sér. 2, **8**, 1879, 195–202.

Dedekind, R.

1. 'Über die Composition der binären quadratischen Formen' (1871), *Ges. Math. Werke*, Bd. III, Braunschweig, 1933, 223–261.
2. *Stetigkeit und Irrationale Zahlen*, Vieweg, Braunschweig, 1872.
3. 'Bernhard Riemann', in: *The Physico-Mathematical Sciences in Past and Present. J. Pure and Appl. Math., Physics, and Astronomy*, **2**, Moscow, 1886, 34–46, 146–158.
4. 'Sur la théorie des nombres entiers algébriques', *Bull. Sci. Math. Astron.*, **11**, 1876, 278–288; sér. 2, **1**, 1877, 17–41, 69–92, 144–164, 207–248. *Ges. Math. Werke*, Bd. III, Braunschweig, 1933, 262–296.
5. *Was sind und was sollen die Zahlen?*, Vieweg, Braunschweig, 1888.
6. *Was sind und was sollen die Zahlen? Stetigkeit und Irrationale Zahlen*, Berlin, 1967.

Dell'Angola, C. A.

1. 'Sopra alcune proposizioni fondamentali dell'analisi', *R. Ist. Lombardo di Sci. e Lett.*, Rend. ser. 2, **40**, 1907, 369–386.

De Moivre, A.

1. *Miscellanea analytica de seribus et quadratures*, Londini, 1730.

Denjoy, A.

1. 'Une extension de l'intégrale de M. Lebesgue', *CR*, **154**, 1912, 859–862.
2. 'Calcul de la primitive de fonction dérivée la plus générale', Ibid., 1075–1078.

Denjoy, A., Felix, L., and Montel, P.

1. 'Henri Lebesgue, le savant, le professeur, l'homme', *L'Enseign. Math.*, ser. 2, **3**, 1957, 1–18.

Descartes, R.
1. *Geometry*, Dover Reprint, 1954.

Dickstein, S.
1. 'Zur Geschichte der Prinzipien der Infinitesimalrechnung. Die Kritiker der "Théorie des fonctions analytiques" von Lagrange', *Abhandl. Geschichte Math.*, **9**, 1899, 65–79.

Dieudonné, J.
1. *Foundations of Modern Analysis*, Academic Press, New York, 1960.
2. 'Histoire de l'analyse harmonique,' XIIIe *Congr. internat. hist. des sci. Colloquium: Voies du développement de l'analyse fonctionnelle*, Moscou, 1971.

Dini, U.
1. 'Sopra una classe di funzioni finite e continue che non hanno mai una derivata', (1877), *Opere Matem.*, V. II, Roma, 1954, 5–7.
2. 'Sulle funzioni limite continue di variabili reali che non hanno mai derivata' (1877), Ibid., 8–11.
3. 'Su alcune funzioni che in tutto un intervallo non hanno mai derivata' (1877), Ibid., 12–31.
4. 'Fondamenti per la teorica delle funzioni di variabili reali', Pisa, 1878.

Dirichlet, J. P. G. Lejeune
1. 'Sur la convergence des séries trigonométriques qui servent à représenter une fonction quelconque dans des limites arbitraires', *J. reine und angew. Math.*, **4**, 1829, 157–169.
2. 'Über die Darstellung ganz willkürlicher Functionen nach Sinus- und Cosinusreihen', (1837), *Werke*, I. Berlin, 1889, 133–160.
3. 'Vorlesungen über die Lehre von den einfachen und mehrfachen bestimmten Integrale', Herausgegeben von G. Arendt, Braunschweig, 1904.

Du Bois-Reymond, P.
1. 'Notiz über einen Cauchy'schen Satz, die Stetigkeit von Summen unendlicher Reihen betreffend', *MA*, **4**, 1871, 135–137.
2. 'Beweis, daß die Koeffizienten der trigonometrischen Reihe $f(x) = \sum_{p=0}^{\infty}(a_p \cos px + b_p \sin px)$ die Werte $a_0 = \frac{1}{2\pi}\int_{-\pi}^{\pi} d\alpha\, f(\alpha)$, $a_p = \frac{1}{\pi}\int_{-\pi}^{\pi} d\alpha\, f(\alpha)\cos p\alpha$, $b_p = \frac{1}{\pi}\int_{-\pi}^{\pi} d\alpha\, f(\alpha)\sin p\alpha$ haben, jedesmal wenn diese Integrale endlich und bestimmt sind' (1874), Leipzig, 1913, 43–91.
3. 'Versuch einer Classification der willkürlichen Functionen reellen Argumentes nach ihren Änderungen in den kleinsten Intervallen', *J. reine und angew. Math.*, **79**, 1875, 21–37.
4. 'Einleitung in der Theorie der bestimmten Integrale von J. Thomae', *Z. Math. Phys. Hist.-liter. Abt.*, **20**, 1875, 121–129.
5. 'Erläuterungen zu den Anfangsgrunden der Variationsrechnung', *MA*, **15**, 283–314.
6. 'Der Beweis des Fundamentalsatzes der Integralrechnung $\int_a^b F'(x)\,dx = F(b) - F(a)$', *MA*, **16**, 1880, 115–127.

7. 'Über den Convergenzgrad der variablen Reihen und Stetigkeitsgrad der Functionen zweier Argumente', *J. reine und angew. Mat.*, **100**, 1887, 331–358.

Dugac, P.
1. 'Eléments d'analyse de Karl Weierstrass', *Arch. for Hist. Exact Sci.*, **10**, 1973, 41–176.
2. 'Notes et documents sur la vie et l'œuvre de René Baire', *Arch. Hist. Exact Sci.*, **15**, No. 4, 1976, 297–383.

Egorov, D. F.
1. 'Sur les suites de fonctions mesurables', *CR*, **152**, 1911, 244–246.

Engels, F.
1. *The Dialectics of Nature*, International Publishers, New York, 1940.

Euler, L.
1. *Introductio in Analysin Infinitorum*, Lausanne, 1748.
2. *Institutiones Calculi Differentialis*, Petersburg, 1755.
3. *Institutionum Calculi Integralis*, T. I, Petersburg, 1768.

Fikhtengol'ts, G. M.
1. 'Definite integrals depending on a parameter' [Russian], *Matem. Sb.*, **29**, 1913, 53–66.
2. 'On change of variables in multiple integrals' [Russian], *IMI*, **5**, 1952, 241–268.

Fischer, E.
1. 'Sur la convergence en moyenne', *CR*, **144**, 1907, 1022–1024.
2. 'Application d'un théorème sur la convergence en moyenne', Ibid., 1148–1151.

Fourier, J. B.
1. *Théorie analytique de Chaleur* (1822), *Œuvres*, T. I, Paris, 1888.
2. 'Mémoire sur la théorie analytique de la chaleur' (1829), *Œuvres*, T. II., Paris, 1890, 145–181.

Fraser, C. G.
1. 'Joseph-Louis Lagrange's algebraic vision of the calculus', *Hist. Math.*, **14**, 1987, 38–53.

Fréchet, M.
1. 'Sur les opérations linéaires', *TAMS*, **5**, 1904, 493–499.
2. 'Sur les fonctions limites et les opérations fonctionnelles', *CR*, **140**, 1905, 27–29.
3. 'La notion d'écart dans le calcul fonctionnel', Ibid., 772–774.
4. 'Sur quelques points du calcul fonctionnel', *Rend. Circolo Mat. Palermo*, **22**, 1906, 1–74.
5. 'Sur l'intégrale d'une fonctionnelle à un ensemble abstrait', *Bull. Soc. Math. France*, **43**, 1915, 248–265.
6. 'Des familles et fonctions additives d'ensembles abstraits', *FM*, **4**, 1923, 329–365; **5**, 1924, 206–251.
7. 'L'analyse générale et les ensembles abstraits', *Rev. de Métaph. et de Morale*, **32**, No. 1, 1925, 1–30.

8. 'L'analyse générale et les espaces abstraits', *Atti Congr. Internat. dei Matematici* (VI), Bologna, 1928, T. 1, Bologna, 1929, 267–274.

Frege, G.

1. 'Funktion und Begriff' (1891), in: *G. Frege. Funktion, Begriff, Bedeutung*, Göttingen, 1962, 16–37.
2. 'Was ist eine Funktion?' (1904), Ibid., 79–88.

Freudenthal, H.

1. 'Did Cauchy plagiarize Bolzano?', *Arch. Hist. Exact Sci.*, **7**, No. 5, 1971, 375–392.

Gaiduk, Yu. M.

1. 'Axel Harnack (1851–1888), graduate of Tartussk University' [Russian], in: *From the History of Science and Technology of the Pribaltic*, No. I(VII), Zinatne, Riga, 1968, 125–132.

Galois, E.

1. 'Remarques sur points diverses d'analyse', (1830–1831), *Œuvres Mathématiques*, Gauthier-Villars, Paris, 1897.

Gauss, K. F.

1. 'Gauss an Schumacher', *Werke*, Bd. X_1, Kgl. Gesellsch. Wiss., Göttingen, 1917, 243–245.

Gerver, J.

1. 'The differentiability of the Riemann function at certain rational multiples of π', *Proc. Nat. Acad. Sci. USA*, **62**, No. 3, 1969, 668–670.
2. 'More on the differentiability of the Riemann function', *Amer. J. Math.*, **93**, 1971, 33–41.

Glivenko, V. I.

1. *The Stieltjes Integral* [Russian], ONTI, Moscow, 1936.
2. 'An experiment in the general definition of the integral' [Russian], *DAN*, **14**, 61–64, 1937.

Gnedenko, B. V.

1. 'The role of the history of physico-mathematical science in the development of modern science' [Russian], *History and Methodology of the Natural Sciences*, V, Moscow University Press, 1966, 5–14.
2. 'On Hilbert's sixth problem' [Russian], in: *The Hilbert Problems*, Nauka, Moscow, 1969, 116–120.

Grabiner, J.

1. *The Origins of Cauchy's Rigorous Calculus*, MIT Press, Cambridge, 1981.

Granger, G. G.

1. *La mathématique social du Marquis de Condorcet*, Paris, 1956.

Grattan-Guinness, I.

1. 'Bolzano, Cauchy, and the "New Analysis" of the early nineteenth century', *Arch. Hist. Exact Sci.*, **6**, No. 5, 1970, 372–400.

2. *The development of the foundations of mathematical analysis from Euler to Riemann*, Cambridge and London, 1970.
3. 'A mathematical union: William Henry and Grace Chisholm Young.' *Annals of Science*, **29**, No. 2, 1972, 105–186.

Gyunter, N. M.

1. 'Sur les intégrales de Stieltjes et leur applications au problème de la physique mathématique', *Trudy Fiz.-matem. Inst. im. Steklova AN SSSR*, **1**, 1932, 1–494.
2. 'La théorie des fonctions de domaines dans la physique mathématique', *Prace Matem.-Fis.*, **44**, 1937, 39–50.

Hadamard, J.

1. 'Sur les opérations fonctionelles', *CR*, **136**, 1903, 351–354.
2. 'Le calcul fonctionnel', *L'Enseign. Math.*, **14**, 1912, 5–18.

Hahn, H.

1. 'Bericht über die Theorie der linearen Integralgleichungen',*Jahresb. Dtsch. Math.-Ver.*, **20**, 1911, 69–117.

Hake, H.

1. 'Über de la Vallée-Poussins Ober- und Unterfunktionen einfacher Integrale und die Integraldefinition von Perron', *MA*, **83**, 1921, 119–142.

Halmos, P. R.

1. *Measure Theory*, Van Nostrand, Princeton, 1950.

Halphen, G. H.

1. 'Sur la série de Fourier', *CR*, **95**, 1882, 1217–1219.

Hankel, H.

1. *Die Entwicklung der Mathematik in dem letzten Jahrhundert*, Tübingen, 1869.
2. 'Untersuchungen über die unendlich oft oszillierenden und unstetigen Functionen' (1870), *Ostwalds Klassiker der Exacten Wissenschaften*, No. 153, Leipzig, 1905, 44–102.

Hardy, G. H.

1. 'The elementary theory of Cauchy's principal values', *PLMS*, **34**, 1901/1902, 16–40.
2. 'The theory of Cauchy's principal values (second paper: The use of principal values in some of the double limit problems of the integral calculus)', Ibid., 55–91.
3. 'The theory of Cauchy's principal values (third paper: Differentiation and integration of principal values)', Ibid., **35**, 1902/1903, 81–107.
4. 'Weierstrass' nondifferentiable function', *TAMS*, **17**, No. 3, 1916, 301–325.
5. 'Sir George Stokes and the concept of uniform convergence', *Proc. Cambridge Phil. Soc.*, **19**, 1918, 148–156.
6. *Divergent Series*, Clarendon Press, Oxford, 1949.
7. 'William Henry Young', *J. London Math. Soc.*, **17**, No. 4, 1942, 218–237.

Harnack, A.

1. 'Über die trigonometrische Reihe und die Darstellung willkürlicher Functionen', *MA*, **17**, 1880, 123–132.
2. 'Vereinfachung der Beweise in der Theorie der Fourier'schen Reihen', Ibid., **19**, 1882, 235–279.
3. 'Berichtigungen zu dem Aufsatze: "Über die Fourier'schen Reihen"', Ibid., 524–238.

Hausdorff, F.

1. *Mengenlehre*, Dritte Auflage, Gruyter, Berlin, 1935.

Hawkins, T.

1. *Lebesgue's Theory of Integration, its Origins and Development*, Madison, Milwaukee, and London, 1970.

Hayes, C. A. and Pauc, C. J.

1. 'Full individual and class differentiation theorems in their relation to halo and Vitali properties', *Can. J. Math.*, **7**, No. 2, 1955, 221–274.

Heine, E.

1. 'Über trigonometrische Reihen', *J. reine und angew. Math.*, **71**, 1870, 335–365.
2. 'Die Elemente der Functionenlehre', Ibid., **74**, 1872, 172–178.

Hilbert, D.

1. *Grundzüge einer allgemeinen Theorie der linearen Integralgleichungen*, Leipzig-Berlin, 1912.

Hilbert's Problems [Russian], Nauka, Moscow, 1969.

Hildebrandt, H.

1. 'The Borel theorem and its applications', *BAMS*, **32**, 1926, 423–474.

Hobson, E. W.

1. 'On modes of convergence of an infinite series of functions of a real variable', *PLMS*, ser. 2, **1**, 1904, 373–387.
2. *The Theory of Functions of a Real Variable and the Theory of Fourier's Series*, Cambridge, 1907.
3. *The Theory of Functions of a Real Variable and the Theory of Fourier's Series*, Vol. II, Cambridge, 1926.

Hoppe, E.

1. 'Zur Geschichte der Infinitesimalrechnung bis Leibniz und Newton', *Jahresber. Dtsch,. Math.-Ver.*, **37**, 1928, 149–187.

Hoüel, J.

1. 'Hankel, H. Untersuchungen über die unendlich oft oscillierenden und unstetigen Functionen. Ein Beitrag zur Feststellung des Begriffs der Function überhaupt.' Universitäts-Programm zum 6 März 1870, Tübingen, *Bull. Sci. math. et astron.*, **1**, 1870, 117–124.

Hugoniot, H.
1. 'Sur le développement des fonctions en séries d'autres fonctions', *CR*, **95**, 1882, 907–909.

Ionescu-Tulcea, C. T.
1. 'Integrale additive', *Com. Acad. Rep. Popul. Romîne*, **4**, No. 9–10, 1954, 471–477.

Jordan, C.
1. 'Sur la série de Fourier (1881)', *Œuvres*, T. IV, Paris, 1964, 393–395.
2. 'Remarques sur les intégrales définies' (1892), Ibid., 428–457.
3. 'Cours d'analyse de l'Ecole polytechnique', 3 Tt., Paris, 1893–1896.

Jourdain, P. E. B.
1. 'Anmerkungen zu Hankel's "Untersuchungen über die unendlich oft oszillierenden und unstetigen Funktionen"', *Ostwald's Klassiker der exacten Wissenschaften*, No. 153, Leipzig, 1905, 103–115.

Kaczmarz, S. and Nikliborc, W.
1. 'Sur les suites convergentes "im Mittel"', *Stud. Math.*, **2**, 1930, 197–206.

Kaczmarz, S. and Steinhaus, H.
1. *Theorie der Orthogonalreihen*, Warsaw, 1935.

Kantorovich, L. V. and Akilov, G. P.
1. *Functional Analysis in Normed Spaces*, Macmillan, New York, 1964.

Kempisty, S.
1. 'Sur les séries itérées des fonctions continues', *FM*, **2**, 1921, 64–73.

Keyser, C. J.
1. 'Three great synonyms: relation, transformation, function,' *Scripta Math.*, **3**, No. 4, 301–316.

Khinchin, A. Ya.
1. 'Studies in the structure of measurable functions' *Matem. Sb.*, **31**, 1924, 265–285, 377–433.

Kitcher, P.
1. 'Fluxions, limits, and infinite littlenesse. A study of Newton's presentations of the calculus', *Isis*, **64**, No. 221, 1973, 33–49.

Kleene, S. C.
1. *Introduction to Metamathetics*, Van Nostrand, Princeton, 1952.

Klein, F.
1. *Vorlesungen über die Entwicklung der Mathematik im 19. Jahrhundert*, Teubner, Leipzig, 1926.

Klement'ev, Z. I. and Bokk, A. A.
1. 'On a generalization of the Sobolev space W_p', [Russian], *Tr. Tomsk. Gos. Univ.*, Ser. Mekh-mat., **179**, 1966, 6–13.

2. 'On the theory of vector-valued measures' [Russian], *Tr. Tomsk. Gos. Univ.*, Ser. Mekh-mat., **179**, 1966, 25–33.

Knopp, K.

1. 'Ein einfaches Verfahren zur Bildung stetiger nirgends differenzierbarer Funktionen', *Math. Z.*, **2**, 1918, 1–26.

Kol'man, E. Ya.

1. *Bernard Bolzano* [Russian], Academy of Sciences of the USSR, Moscow, 1955.

Kolmogorov, A. N.

1. 'Une série de Fourier-Lebesgue divergente partout', *CR*, **183**, 1926, 1327–1328.
2. 'Untersuchungen über den Integralbegriff', *MA*, **103**, No. 3–4, 1930, 654–696.
3. 'Newton and modern mathematical thought' [Russian], in: *Moscow University - in Memoriam, Isaac Newton*, Moscow University Press, 1946, 27–42.

Kolmogorov, A. N. and Fomin, S. V.

1. *Elements of the Theory of Functions and Functional Analysis*, Graylock Press, New York, 1957.

König, J.

1. 'A hatarozott integralok elmeletéhez', *Mat. és. Természettudományi értesítő*, 1897, 380–384.

Korovkin, P. P.

1. 'A generalization of Egorov's Theorem' [Russian], *DAN*, **58**, No. 7, 1947, 1265–1267.

Krasnosel'skii, M. A. and Rutickii, Ya. B.

1. *Convex Functions and Orlicz Spaces*, Noordhoff, Groningen, 1961.

Kudryavtsev, L. D. and Nikol'skii', S. M.

1. 'Theory of differentiable functions of several variables (embedding theorems)' [Russian], in: *History of Russian and Soviet Mathematics* [Russian], T. 3, Academies of Sciences of the USSR and the Ukrainian SSR, Kiev, 1968, 588–607.

Kuratowski, K.

1. 'Les fonctions semicontinues dans l'espace des ensembles fermés', *FM*, **18**, 1932, 148–159.
2. *Topology*, Vol. 1, Academic Press, New York, 1966.

Kuratowski, K. and Mostowski, A.

1. *Set Theory*, North-Holland Publishing Company, Amsterdam, 1968.

Kuznetsov, B. G.

1. *Von Galilei bis Einstein: Entwicklung der Physicalischen Ideen*, C. F. Winter, Basel, 1970.

Lacroix, S. F.

1. *Traité du calcul différentiel et du calcul intégral*, I–II, Paris 1797–1798.

Lagrange, J. L.
1. 'Sur une nouvelle espèce de calcul relatif à l'intégration des quantités variables' (1772), *Œuvres*, T. III, Paris, 1869, 441–476.
2. 'Théorie des fonctions analytiques, contenant les principes du calcul différential, dégagés de toute considération d'infiniment petits, d'évanouissants, de limites et de fluxions, et réduites à l'analyse algébrique des quantités finis' ($1^{\text{ère}}$ éd., Paris, 1797; $2^{\text{ième}}$ éd., 1813), *Œuvres*, T. IX, Paris, 1881, 1–428.
3. *Leçons sur le calcul des fonctions* (1801), *Œuvres*, T. 10, Paris, 1884.

Landau, E.
1. 'Ein Satz über Riemannsche Integrale', *Math. Z.*, **2**, 1918, 350–351.

Lavrent'ev, M. A. and Shabat, B. V.
1. *Complex Variable Methods* [Russian], Moscow-Leningrad, Gostekhizdat, 1951.

Lebesgue, H.
1. 'Sur les fonctions de plusieurs variables', *CR*, **128**, 1899, 811–813.
2. 'Sur une généralisation d'intégrale définie', *Ibid.*, **132**, 1901, 1025–1028.
3. 'Intégrale, longueur, aire', *Ann. Matem. Pura ed Appl.*, ser. 3, **17**, 1902, 231–359.
4. 'Sur l'existence des dérivées', *CR*, **136**, 1903, 659–661.
5. 'Sur une propriété des fonctions', *Ibid.*, **137**, 1903, 1228–1230.
6. 'Sur les séries trigonométriques', *Ann. Sci. de l'Ec. Norm. Sup.*, sér. 3, **20**, 1903, 453–485.
7. 'Une propriété caractéristique des fonctions de classe 1', *Bull. Soc. Math. France*, **32**, 1904, 229–242.
8. *Leçons sur l'intégration et la recherche des fonctions primitives*, Paris, 1904.
9. 'Sur les fonctions représentables analytiquement', *CR*, **139**, 1904, 29–31.
10. 'Sur les fonctions représentables analytiquement', *J. de Math. Pures et Appl.*, sér. 6, **1**, 1905, 139–216.
11. 'Remarques sur la définition de l'intégrale', *Bull. des Sci. Math.*, sér. 2, **29**, 1905, 272–275.
12. *Leçons sur les séries trigonométriques*, Paris, 1906.
13. 'Sur la recherche des fonctions primitives par l'intégration', *Atti della R. Accad. dei Lincei*, Cl. delle sci. fis., matem. e nat., ser. 5, **161**, 1907, 283–290.
14. 'Sur les intégrales singulières', *Ann. Fac. Sci. l'Univ. Toulouse*, sér. 3, **1**, 1909, 25–117.
15. 'Sur l'intégrale de Stieltjes et sur les opérations fonctionnelles linéaires', *CR*, **150**, 1910, 86–88.
16. 'Sur l'intégration des fonctions discontinues', *Ann. Sci. de l'Ec. Norm. Sup.*, sér. 3, **27**, 1910, 361–450.
17. 'Sur certaines démonstrations d'existence', *Bull. Soc. Math. Fr.*, **45**, 1917, 132–144.
18. 'Remarques sur les théories de la mesure et de l'intégration', *Ibid.*, sér. 3, **35**, 1918, 191–250.
19. 'Sur une définition due à M. Borel', *Ibid.*, sér. 3, **37**, 1920, 255–257.
20. *Leçons sur l'intégration et la recherche des fonctions primitives*, deuxième éd., Gauthier-Villars, Paris, 1928.

Leibniz, G. W.

1. 'Nova methodus pro Maximis et Minimis, itemque Tangentibus, quae nec fractas, nec irrationales quantitates moratur, et singulis pro illis calculi genus', *Acta Erud.*, 1684, in: *Mathematische Schriften*, Bd. V, Georg Olms Verlag, New York, 1971, 220–225.

Lobachevskii, N. I.

1. 'On the vanishing of trigonometric series' (1834), *Works* [Russian], T. 5, Gostekhizdat, Moscow-Leningrad, 1951, 31–80.
2. 'A method of verifying the vanishing of infinite series and approximating the value of functions of very large numbers' (1835), Ibid., 81–162.

Looman, H.

1. 'Über die Perronsche Integraldefinition', *MA*, **93**, 1924, 153–156.

Lunts, G. L.

1. 'On the works of N. I. Lobachevskii in mathematical analysis' [Russian], *IMI*, **2**, 1949, 9–71.
2. 'A survey of the works of N. I. Lobachevskii' [Russian], in: *N. I. Lobachevskii: Collected Works*, T. 5, Gostekhizdat, Moscow-Leningrad, 1951, 13–30.

Luzin, N. N.

1. 'On the fundamental theorem of integral calculus' [Russian] (1912), *Collected Works*, T. I, Academy of Sciences of the USSR, Moscow, 1953, 5–24.
2. 'Sur les propriétés fondamentales des fonctions mesurables', *CR*, **154**, 1912, 1688–1690.
3. 'Sur un problème de M. Baire', *CR*, **158**, 1914, 1258–1261.
4. 'The integral and the trigonometric series' [Russian] (1915), *Collected Works* [Russian]., T. I, 1953, 48–212.
5. 'Sur l'existence d'un ensemble non-dénombrable qui est de première catégorie dans tout ensemble parfait' *FM*, **2**, 1921, 155–157.
6. *Leçons sur les ensembles analytiques et leurs applications*, Gauthier-Villars, Paris, 1930.
7. 'The current state of the theory of functions of a real variable' [Russian] (1933), *Collected Works* [Russian], T. II, Academy of Sciences of the USSR, Moscow, 1958, 494–536.
8. 'Euler (1707–1783)' [Russian], Ibid., T. III, 1959, 351–372.
9. *The Theory of Functions of a Real Variable* [Russian], Uchpedgiz, Moscow, 1948.

Luzin, N. N. and Sierpiński, W.

1. 'On an uncountable set that is of first category on every perfect set', *Rend. Accad. Lincei*, Ser. 6, **7**, No. 3, 1928, 214–215; also in: *N. N. Luzin: Collected Works* [Russian], T. II, Academy of Sciences of the USSR, Moscow, 1958, 697–698.

Mac Laurin, C.

1. *A treatise of Fluxions*, Edinbourgh, 1742.

Maeda, F.

1. 'On the definition and the approximate continuity of the general derivate', *J. Sci. Hiroshima University*, ser. A, **2**, No. 1, 1932, 33–53.

Maistrov, L. E.
1. *Probability Theory*, Academic Press, New York, 1974.

Mal'cev, A. I.
1. *Algebraic Systems*, Springer, New York, 1973.

Marcus, S.
1. 'Quelques aspects des travaux roumaine dans la théorie des fonctions de variables réelles,' *Rev. Roumaine de Math. Pures et Appl.*, **9**, No. 9, 1966, 1075–1102.

Markushevich, A. I.
1. *Essays on Analytic Function Theory* [Russian], Gostekhizdat, Moscow-Leningrad, 1951.
2. 'The works of Gauss on mathematical analysis' [Russian], in: *Karl Friedrich Gauss*, Academy of Sciences of the USSR, Moscow, 1956, 145–216.

Mazurkiewicz, L. L.
1. *Considérations générales sur les principes d'analyse infinitésimale suivés d'exposé de l'intégration directe indépendante du calcul différentiel*, St.-Pétersbourg, 1875.

Mazurkiewicz, S.
1. 'Sur les fonctions de classe 1', *FM*, **2**, 1921, 28–36.
2. 'Sur les fonctions non dérivables', *Stud. Math.*, **3**, 1931, 174–179.

Medvedev, F. A.
1. 'On Cauchy's coexisting quantities' [Russian], in: *History of the Physico-Mathematical Sciences*, Academy of Sciences of the USSR, Moscow, 1961, 264–289.
2. *The Development of Set Theory in the Nineteenth Century* [Russian], Nauka, Moscow, 1965.
3. 'Set functions in the writings of G. Peano' [Russian], *IMI*, **16**, 1965, 311–323; slightly abbreviated translation: Medvedev, F. A., 'Le funzioni d'insieme secondo G. Peano, *Arch. Intern. d'Hist. des Sci.*, **33**, No. 10, 1983, 112–117.
4. 'Les quadratures et les cubatures chez Pappus d'Alexandrie', *Actes de XIIe Congrés Intern. d'Hist. de Sci.*, T. IV, Paris, 1971, 107–110.
5. *The French School of the Theory of Functions and Sets at the Turn of the Twentieth Century* [Russian], Nauka, Moscow, 1976.

Menger, K.
1. 'Analytische Funktionen', in: *Festschrift zur Gedächtnisfeier für Karl Weierstrass, 1815–1965*, Köln and Oplanden, 1966, 609–612.

Men'shov, D. E.
1. 'Sur l'unicité du développement trigonométrique', *CR*, **163**, 1916, 433–436.
2. 'Sur la représentation des fonctions mesurables par des séries trigonométriques', *Matem. Sb.*, **9(51)**, 1941, 667–692.
3. 'On the convergence in measure of trigonometric series' [Russian], *Tr. Matem. Inst. im. V. A. Steklova AN SSSR*, **32**, 1950.

Montel, P.
1. 'Sur les suites infinies de fonctions', *Ann. Sci. de l'Ec. Norm. Sup.*, sér. 3, **24**, 1907, 233–334.

Moore, E. H.
1. 'The definition of limit in general analysis', *Proc. Nat. Acad. Sci. USA*, ser. A, **1**, No. 12, 1915, 628–632.

Moore, E. H. and Smith, H. L.
1. 'A general theory of limits', *Amer. J. Math.*, **44**, No. 2, 1922, 102–121.

Nalli, P.
1. 'Sopra una nuova specie di convergenze in media', *Rend. Circolo Matem. di Palermo*, **38**, 1914, 315–319.
2. 'Aggiunta alla memoria: "Sopra una nuova specie di convergenze in media"', Ibid., 320–323.
3. 'Sopra un'applicazione della convergenze in media, note I, II', *Atti della R. Accad. dei Lincei*, Rend., Cl. delle Sci. Fis., Matem. e Nat., ser. 5, **25**, 1 sem., 1916, 149–155, 284–289.

Natanson, I. P.
1. *Theory of Functions of a Real Variable*, Ungar, New York, 1961.

Natucci, A.
1. 'Saggio storico sulla teoria delle funzioni', *G. Matem. Battaglini*, ser. 5, **87**, 1959, 89–146.

Newton, I.
1. 'De analysi per æquationes infinitas', *Mathematical Papers*, Vol. II, Cambridge University Press, 1967, 206–247.
2. *The Method of Fluxions and Infinite Series*. Henry Woodfall, London, 1736.
3. 'Letter to Oldenburg', *Correspondence of Isaac Newton*, Vol. II, Cambridge University Press, 1960, 110–161.
4. 'De Quadratura Curvarum', *Mathematical Papers*, Vol. VII, 1976, 24–182.
5. *Mathematical Principles of Natural Philosophy*, University of California Press, Los Angeles, 1966.

Nikodým, O.
1. 'Sur une généralisation des intégrales de M. J. Radon', *FM*, **15**, 1930, 131–179.

Nikol'skii, S. M.
1. 'Some properties of differentiable functions defined on an n-dimensional open set' [Russian], *IAN*, **23**, 1959, 213–242.
2. *Approximation of Functions of Several Variables and Imbedding Theorems*, Springer, New York, 1975.

Orlicz, W.
1. 'Nonuniform convergence and the integration of series term by term', *Amer. J. Math.*, **19**, No. 2, 1897, 155–190.

Osgood, W. F.

1. 'Sur les fonctions continues non-dérivable', *FM*, **34**, 1917, 45–60.

Ostrogradskii, M. V.

1. 'An excerpt from the second public lecture on transcendental analysis read by Academician Ostrogradskii' (1841), *Collected Works* [Russian], T. III, Academy of Sciences of the Ukrainian SSR, Kiev, 1961, 165–170.

Oxtoby, J. C.

1. *Measure and Category*, Springer, New York, 1971.

Paplauskas, A. B.

1. *Trigonometric Series from Euler to Lebesgue* [Russian], Nauka, Moscow, 1966.

Pascal, E.

1. *Esercizi e note critiche di calcolo infinitesimale*, Milano, 1895.

Pasch, M.

1. *Einleitung in die Differential- und Integralrechnung*, Leipzig, 1882.
2. 'Über einige Punkte der Funktionenlehre', *MA*, **30**, 1887, 132–154.
3. *Veränderliche und Funktion*, Leipzig u. Berlin, 1914.

Peano, G.

1. *Applicazioni geometriche del calcolo infinitesimale*, Torino, 1887.
2. 'Sur une courbe qui remplit tout une aire plane' (1890), *Opere Scelte*, T. I, Roma, 1957, 110–114.
3. 'Sulla definizione di integrale' (1895), Ibid., 277–281.
4. 'Sulla definizione di funzione' (1911), Ibid., 363–365.
5. 'Le grandezze coesistenti di Cauchy' (1914), Ibid., 432–440.

Perron, O.

1. 'Über den Integralbegriff', *Sitzungsber. Heidelb. Akad. Wiss.*, **14**, 1914.

Pesin, I. N.

1. *The Development of the Concept of Integral* [Russian], Nauka, Moscow, 1966.

Pincherle, S.

1. 'Sopra alcuni sviluppi in serie per funzioni analitiche' (1882), *Opere Scelte*, I, Roma, 1954, 64–91.
2. 'Mémoire sur le calcul fonctionnel distributif' (1897), Ibid., II, Roma, 1954, 1–70.

Plancherel, M.

1. 'Contribution à l'étude de la représentation d'une fonction arbitraire par des intégrales définies', *Rend. Circolo Matem. Palermo*, **30**, 1910, 289–336.

Pogrebysskii, I. B.

1. *From Lagrange to Einstein. The Classical Mechanics of the Nineteenth Century* [Russian], Nauka, Moscow, 1966.
2. *Gottfried Wilhelm Leibniz* [Russian], Nauka, Moscow, 1971.

Pringsheim, A.
1. 'Über die notwendigen und hinreichenden Bedingungen des Taylor'schen Lehrsatz für Funktionen einer reellen Variablen', *MA*, **44**, 1894, 57–82.
2. 'Grundlagen der allgemeinen Funktionenlehre', *Enzykl. math. Wiss. mit Einschluss ihrer Anwendungen*, Bd. II, **A** 1, Leipzig, 1899–1916, 1–53.

Privalov, I. I.
1. 'On Vitali's theorem and a result of Prof. Steklov' [Russian], *Matem. Sb.*, **30**, No. 3, 1916, 295–298.

Raabe, J. L.
1. *Die Differential- und Integralrechnung mit Functionen einer Variabeln*, Erster Teil. Zürich, 1839.
2. 'Über den Fall, wenn in dem bestimmten Integrale $\int_a^b \varphi(x)\,dx$ die Function $\varphi(x)$ für einen oder mehrere Werthe von x, welche innerhalb a und b liegen, undendlich gross oder discontinuirlich wird', *J. reine und angew. Math.*, **20**, 1839, 173–177.

Radon, J.
1. 'Theorie und Anwendungen der absolut additiven Mengenfunktionen', *Sitzungsber. K. Akad. Wiss. Wien*, Math.-naturwiss. Kl., Abt., IIa, **122**, No. 7, 1913, 1295–1438.
2. 'Über lineare Funktionaltransformationen und Funktionalgleichungen', *Sitz. der Akad. der Wiss. Wien*, **128**, 1919, 1083–1121.

Remez, E. Ya.
1. 'On the mathematical manuscripts of Academician M. V. Ostrogradskii' [Russian], *IMI*, **4**, 1959, 9–98.

Reyneaux, Ch. R.
1. *Usage de l'analyse, ou la manière de l'appliquer à decouvrir les propriétés des figures de la géometrie simple & composée, à resoudre les Problèmes de ces sciences & les Problèmes des sciences Physico-mathématiques, en employant le calcul ordinaire de l'Algèbre, le calcul différentiel & le calcul intégral. Ces derniers calculs sont aussi expliqués & démontrés*, Second éd., augmentée des Remarques de M. de Varignon, T. II, Paris, MDCCXXXVIII.

Riemann, B.
1. 'Grundlagen für eine allgemeine Theorie der Functionen einer veränderlichen komplexen Grösse' (1851), *Ges. Abh.*, Dover, New York, 1953, 1–48.
2. 'Über die Darstellbarkeit einer Function durch eine trigonometrische Reihe', Ibid., 227–271.

Riesz, F.
1. 'Sur les ensembles de fonctions' (1906), *Œuvres Complètes*, T. I, Budapest, 1960, 375–377.
2. 'Sur les systèmes orthogonaux de fonctions et l'équations de Fredholm' (1907), Ibid., 382–385.
3. 'Über orthogonale Funktionensysteme' (1907), Ibid., 385–395.

4. 'Sur les systémes orthogonaux de fonctions' (1907), Ibid., 615–619.
5. 'Sur les suites de fonctions mesurables' (1909), Ibid., 396–399.
6. 'Sur les opérations fonctionnelles linéaires' (1909), Ibid., 400–402.
7. 'Untersuchungen über Systeme integrierbarer Funktionen' (1910), Ibid., 441–489.
8. 'Sur certains systèmes singuliers d'équations intégrales' (1911), Ibid., T. II, 798–827.
9. 'Über Integration unendlicher Folgen' (1918), Ibid., 195–199.
10. 'Sur le théorème de M. Egoroff et sur les opérations fonctionnelles linéaires', (1922–23), Ibid., 215–226.
11. 'Sur l'intégrale de Lebesgue' (1920), Ibid., 200–214.
12. 'Sur l'existence de la dérivée des fonctions monotones et sur quelques problèmes qui s'y rattachent' (1932), Ibid., 250–263.

Ringenberg, L. A.
1. 'The theory of the Burkill integral', *Duke Math. J.*, **15**, 1948, 239–270.

Rosenthal, A.
1. 'Neuere Untersuchungen über Funktionen reeller Veränderlichen', Leipzig-Berlin, 1924.

Rozhanskaya, M. M.
1. 'On functional dependencies in the "al-Qānūnū'l-Mas'ūdī" of al-Bīrūnī' [Russian], *Vestn. Kara-Kalpak filiala AN UzbSSR*, (Messenger of the Kara-Kalpak branch of the Uzbek Academy of Sciences), ser. math., **4**, 1966, 14–22.
2. 'Methods of studying the general properties of functions in the "al-Qānūnū'l-Mas'ūdī" of al-Bīrūnī', Ibid., **1**, 1967, 29–35.
3. 'Functional dependencies in the writings of al-Bīrūnī', Abstract of Kandidat Dissertation, Moscow, 1967.

Russell, B.
1. *The Principles of Mathematics*, W. W. Norton & Company, New York, 1938.
2. 'Sur la relation des mathématiques à la logistique', *Rev. métaph. morale*, **13**, 1905, 906–916.

Rybnikov, K. A.
1. *History of Mathematics* [Russian], T. II, Moscow Univ. Press, 1963.

Saks, S.
1. 'On the functions of Besicovitch in the space of continuous functions', *FM*, **19**, 1932, 211–219.
2. *Theory of the Integral*, Stechert, New York, 1937.

Schmidt, E.
1. 'Sur la puissance des systèmes orthogonaux de fonctions continues', *CR*, **143**, 1906, 955–957.

Schönflies, A.
1. 'Die Entwicklung der Lehre von den Punktmannigfaltigkeiten. Bericht, erstattet der Deutschen Mathematiker-Vereinigung', *Jahresb. Dtsch. Math.-Ver.*, **8**, 1900, 1–251.

2. 'Die Entwicklung der Lehre von den Punktmannigfaltigkeiten', *Jahresb. Dtsch. Math. Ver.*, Ergsbd., **2**, 1908.

Schramm, M.

1. 'Steps towards the idea of function: a comparison between eastern and western science of the middle ages. Some further remarks on A. Crombie's *Augustine to Galileo*', *History of Science*, **4**, 1956, 70–103.

Schröder, E.

1. *Algebra und Logik der Relative*, Leipzig, 1895.

Schwarz, H. A.

1. 'Beispiel einer stetigen nicht differentiirbaren Function' (1873), *Ges. Math. Abh.*, Bd. II, Berlin, 1890, 269–274.

Scriba, Ch. J.

1. *James Gregory's frühe Schriften zur Infinitesimalrechnung*, Giessen, 1957.

Seidel, Ph. L.

1. 'Note über eine Eigenschaft der Reihen, welche discontinuirliche Functionen darstellen' (1847), *Ostwald's Klassiker der exacten Wissenschaften*, No. 116, Leipzig, 1900, 35–45.

Severini, C.

1. 'Sopra gli sviluppi in serie di funzioni orthogonali', *Atti della Accad. Gioenia di Sci. Nat. in Catania*, Anno LXXXVII, ser. 5, **III**, memoria XI, 1910, 1–7.
2. 'Sulle successioni di funzioni ortogonali', *Ibid.*, memoria, XIII, 1–10.

Shatunovskii, S. O.

1. *Introduction to Analysis* [Russian], Odessa, 1923.

Shchegol'kov, E. A.

1. 'Elements of the theory of B-sets', *UMN*, **5**, No. 5(39), 1950, 14–44.

Shikhanovich, Yu. A.

1. *Introduction to Modern Mathematics* [Russian], Nauka, Moscow, 1965.

Shilov, G. E.

1. *Analyse Mathématique*, 1ère et 3ème partie, Mir, Moscow, 1973.

Sierpiński, W.

1. 'Sur les fonctions développables en séries absolument convergentes des fonction continues', *FM*, **2**, 1921, 15–27.

Sikorski, R.

1. *Boolean Algebras*, Third edition, Springer, Berlin, 1969.

Sirvint, Yu. F.

1. 'Weak compactness in Banach spaces' [Russian], *DAN*, **28**, No. 3, 1940, 199–201.

Smirnov, V. I.
1. *A Course of Higher Mathematics* [Russian], T. 5, Fizmatgiz, Moscow, 1959.

Smirnov, V. I. and Sobolev, S. L.
1. 'N. M. Gyunter', *Uch. Zapiski Leningrad. Univ.*, **15**, 1948, 5–22.

Smith, H. J. S.
1. 'On the integration of discontinuous functions', *PLMS*, **6**, 1875, 140–153.

Sobolev, S. L.
1. *Some applications of functional analysis in mathematical physics* [Russian], Leningrad University Press, 1950.

Steiner, H.-G.
1. 'Aus der Geschichte des Funktionsbegriffs', *Mathematikunterricht*, **15**, No. 3, 1969, 13–39.

Steklov, V. A.
1. 'Sur certaines égalités remarquables', *CR*, **135**, 1902, 783–786.

Stieltjes, T. J.
1. 'Recherches sur les fractions continues', *Toulouse Fac. Sci. Ann.*, **8**, 1894; **9**, 1895.

Stokes, G. G.
1. 'On the critical values of the sums of periodic series' (1848), *Mathematical and Physical Papers*, V. I, Cambridge, 1880, 236–313.

Stolz, O.
1. 'B. Bolzano's Bedeutung in der Geschichte der Infinitesimalrechnung', *MA*, **18**, 1881, 256–279.
2. *Vorlesungen über allgemeine Arithmetik*, Bd. I–II, Leipzig, 1885.

Styazhkin, N. I.
1. *The Creation of Mathematical Logic* [Russian], Nauka, Moscow, 1967.

Talalyan, A. A.
1. 'The representation of measurable functions by series' [Russian], *UMN*, **15**, No. 5, 1960, 77–141.
2. 'On the limiting functions of series with respect to bases in L_p' [Russian], *Matem. Sb.*, **56(98)**, 1962, 353–374.

Tarski, A.
1. *Introduction to Logic and to the Methodology of Deductive Sciences*, Third Edition, Revised, Oxford University Press, New York, 1965.

Thomae, J.
1. *Einleitung in die Theorie der bestimmten Integrale*, Halle a/S, 1875.

Thomé, L. W.
1. 'Über die Kettenbruchentwicklung der Gaußschen Function $F(\alpha, 1, \gamma, x)$', *J. reine und angew. Math.*, **66**, 1866, 322–336.

Timchenko, I. Yu.

1. 'Foundations of the theory of analytic functions' [Russian], *Zapiski Matem. Otd. Novorossiisk. Obshch. Estestvoisp.*, **XII**, 1892, 1–256; **XVI**, 1896, 1–216, [257–472]; **XIX**, 1899, 1–183, [473–655]. Separate edition with the same title, Odessa, 1899.

Tolstov, G. P.

1. *Fourier Series*, Prentice-Hall, Englewood Cliffs, 1962.

Ul'yanov, P. L.

1. 'Theory of functions of a real variable', in: *History of Russian and Soviet Mathematics* [Russian], T. 3, Academies of Sciences of the USSR and the Ukrainian SSR, 1968, 530–568.
2. 'Representation of functions by series and the classes $\varphi(L)$' [Russian], *UMN*, **27**, No. 2(164), 1972, 3–52.

Uspenskii, V. A.

1. 'The contribution of N. N. Luzin to descriptive set theory and descriptive function theory: concepts, problems, predictions' [Russian], *UMN*, **40**, No. 3(243), 1985, 85–116.

Vallée-Poussin, Ch. de la

1. *Cours d'analyse infinitesimale*, Tt. I, II, Paris
2. *Intégrales de Lebesgue. Fonctions d'ensemble. Classes de Baire*, Paris, 1916.

Vinogradova, I. A.

1. 'On the indefinite A-integral' [Russian], *IAN*, **25**, No. 1, 1961, 113–142.
2. 'On the representation of a measurable function by the indefinite A-integral' [Russian], Ibid., **26**, 1962, 581–604.

Vinogradova, I. A. and Skvortsov, V. A.

1. 'Generalized Fourier series and integrals' [Russian], in: *Mathematical Analysis* (1970), Nauka, Moscow, 1971, 65–107.

Vitali, G.

1. 'Sulle funzioni integrali', *Atti della R. Accad. delle sci. di Torino*, **40**, 1904/1905, 1021–1034.
2. 'Sui gruppi di punti e sulle funzioni di variabili reali', *Atti della R. Accad. delle sci. di Torino*, **43**, 229–246.

Volterra, V.

1. 'Alcune osservazioni sulle funzioni punteggiate discontinue', (1881), *Opere matem.*, V. I, Roma, 1954, 7–15.
2. 'Sui principii del calcolo integrale', (1881), Ibid., 16–48.
3. 'Sui fondamenti della teoria delle equazioni differenziali lineari' (1887), Ibid., 209–290.

Weierstrass, K.

1. 'Zur Functionenlehre' (1880), *Mathematische Werke*, Bd. II, Berlin, 1895, 201–223.

2. 'Über die analytischer Darstellbarkeit sogenannter willkürlicher Functionen reeler Argument' (1885), *Mathematische Werke*, Bd. III, Berlin, 1903, 1–37.
3. 'Zur Theorie der Potenzreihen' (1894), *Mathematische Werke*, Bd. I, Berlin, 1894, 67–74.
4. 'Definition analytischer Functionen einer Varänderlichen vermittelst algebraischer Differentialgleichungen' (1894), Ibid., 75–84.
5. 'Über continuirliche Functionen eines reellen Arguments die für keine Werth des letzteren einen bestimmten Differentialquotienten besitzen' (1872), *Mathematische Werke*, Bd. II, Berlin, 71–74.

Weyl, H.
1. 'Über die Konvergenz von Reihen, die nach Orthogonalfunktionen fortschreiten', *MA*, **67**, 1909, 225–245.

Wieleitner, H.
1. *Geschichte der Mathematik*, Part II, from Descartes to about 1800, 2 Vols., Leipzig, 1911–1921.

Wiener, N.
1. *I am a Mathematician*, Doubleday, Garden City, 1956.

Young, R. C.
1. 'On many-sided Riemann-Stieltjes integration, I–II', *Proc. Cambridge Phil. Soc.*, **27**, No. 3, 1931, 326–380.

Young, W. H.
1. 'On nonuniform convergence and the integration of series term-by-term', *PLMS*, ser. 2, **1**, 1904, 89–102.
2. 'Open sets and the theory of content', *PLMS*, ser. 2, **2**, 1905, 16–51.
3. 'On a new method in the theory of integration', Ibid., **9**, 1911, 15–50.
4. 'On some classes of summable functions and their Fourier series', *Proc. R. Soc. of London*, **A 87**, 1912, 225–229.
5. 'On the integration with respect to a function of bounded variation', *PLMS*, ser. 2, **13**, 1914, 109–150.
6. 'On integrals and derivatives with respect to a function', Ibid., **15**, 1916, 35–63.
7. 'On successions with subsequences converging to an integral', Ibid., **24**, 1926, 1–20.

Yushkevich, A. P., Ed.
History of Mathematics from Ancient Times to the Beginning of the Nineteenth Century [Russian], 3 vols., Nauka, Moscow, 1970–1972.
1. *From Ancient Times to the Beginning of the Modern Era*, 1970.
2. *The Mathematics of the Seventeenth Century*, 1970.
3. *the Mathematics of the Eighteenth Century*, 1972.

Yushkevich, A. P.
1. 'The ideas on the foundation of analysis in the eighteenth century' [Russian], in: L. Carnot. *"Reflexions sur la métaphysique du calcul des infiniment petits"* [Russian], GTTI, Moscow-Leningrad, 1933, 7–57.

2. 'On the origin of Cauchy's concept of a definite integral' [Russian], *Tr. Inst. Ist. Estestvoznaniya*, **1**, Academy of Sciences of the USSR, Moscow, 1956, 373–411.
3. 'On Thābit ibn-Qurra's quadrature of the parabola' [Russian], in: *History and Methodology of the Natural Sciences* [Russian], Moscow University Press, **5**, 1966, 118–125.
4. 'On the development of the concept of a function' [Russian], *IMI*, **17**, 1966, 123–150.
5. *History of Mathematics in Russia* [Russian], Nauka, Moscow, 1968.
6. 'On the revolution in modern mathematics' [Russian], *Vopr. Ist. Estestvozn. i Tekh.*, **2**, No. 27, 1969, 14–22; **4**, No. 33, 1971, 3–13.
7. 'The concept of a function in the writings of Condorcet' [Russian], *IMI*, **19**, 1974, 158–166.

Zeller, K.
1. *Theorie der Limitierungsverfahren*, Berlin, Göttingen, Heidelberg, 1958.

Zermelo, E.
1. 'Beweis, daß jede Menge wohlgeordnet werden kann', *MA*, **59**, 1904, 514–516.

Zeuthen, H. G.
1. *Geschichte der Mathematik im 16. und 17. Jahrhundert* (1903), Johnson Reprint Corporation, New York, 1966.

Zubov, V. P.
1. *The development of atomistic conceptions to the beginning of the nineteenth century* [Russian], Nauka, Moscow, 1965.

Index of names

Abel, N. H. (1802–1829) 87, 88, 90, 93, 185, 235
Agnesi, M. G. (1718?–1799) 41, 235
Agostini, A. (b. 1892) 175, 235
Akilov, G. P. (b. 1921) 25, 156, 229, 246
Al-Bīrūnī (973–ca. 1060) 30, 172, 254
Al-Haitham (965–1039) 171
Al-Kuhi (Tenth century) 171
Aleksandrov, P. S. (1896–1986) 25, 110, 197, 229, 235
Ampère, A. M. (1775–1836) 214–219, 232, 235
Antropova, V. A. (b. 1924) 64, 182, 236
Apollonius (ca. 260–170 B. C. E.) 30, 31
Arbogast, L. F. A. (1759–1803) 65, 191
Archimedes, (287–212 B. C. E.) 30, 31, 170–173, 237
Arendt, G. (1832–?) 188, 241
Arzelà, C. (1847–1912) 102–111, 120, 121, 157, 166, 236
Ascoli, G. (1843–1896) 104, 109, 190, 236
Averbuch, V. I. (b. 1937) 196, 236

Baire, R. L. (1874–1932) 13, 17, 20, 106, 112, 115, 125, 157–167, 190, 229–230, 236
Banach, S. (1892–1945) 14, 66, 155, 156, 231–233, 236
Bari, N. K. (1901–1961) 24, 114, 229, 236, 237
Baron, M. E. 179, 237
Barrow, I. (1630–1677) 29, 32, 40, 175, 176
Bashmakova, I. G. (b. 1921) 42, 170, 237
Bauer, H. (1891–1953) 197, 237
Bendixson, I. (1861–1935) 102, 103, 106, 237
Bernkopf, M. 138, 145, 237
Bernoulli, D. (1700–1782) 46, 47
Bernoulli, James (1654–1705) 40

Bernoulli, John (1667–1748) 29, 34, 40–42, 47, 48, 64
Bernshtein, S. N. (1880–1968) 10, 83
Bertrand, J. (1822–1900) 220
Besicovitch, A. S. (Bezikovich, A. S.) (1891–1970) 231, 237, 254
Bessel, F. W. (1784–1846) 83
Bezikovich A. S. (Besicovitch, A. S.) (1891–1970) 231, 237, 254,
Bieberbach, L. (1896–1982) 121, 237
Birkoff, G. (b. 1911) 196
Birnbaum, Z. W. 152–156, 237
Biryukov, B. V. (b. 1922) 70–72, 74, 237
Bochner, S. (1899–1982) 28, 177, 196, 237
Bokk, A. A. (b. 1939) 210, 246
Bolzano, B. (1781–1848) 50, 56, 62, 65, 186, 216, 220, 238, 243, 247
Boole, G. (1815–1864) 70, 205
Borel, E. (1871–1956) 13, 20, 23, 76, 87, 97–98, 125–126, 133, 162–164, 193, 194, 238, 248
Bortolotti, E. (1866–1947) 175, 176, 238
Bouquet, J. C. (1819–1885) 90
Bourbaki, N. 15, 22, 26, 31, 68, 176, 238
Boussinesq, J. V. (1842–1929) 233
Boyer, C. B. (b. 1906) 179, 238
Bradwardine, T. (ca. 1290–1349) 31, 172
Braun, Ben-Ami 128, 238
Briot, Ch. (1817–1882) 90
Brunschvicg, L. A. (1869–1944) 65–66, 212, 219, 238
Bržečka, V. F. 65, 238
Bunyakovskii, B. Ya. (1804–1889) 57
Burkill, J. C. (b. 1900) 154–155, 196, 202, 238, 254

Cahen, E. 224, 239
Cantor, G. (1845–1918) 14, 22, 67–69, 72, 75, 92–95, 111, 132, 160–163, 192, 229, 239

Cantor, M. (1829–1920) 29, 239
Carathéodory, C. (1873–1950) 23, 80, 81, 119, 205–209, 209, 210, 229, 239
Carnot, L. N. M. (1753–1823) 44, 239
Cauchy, A. L. (1789–1857) 14, 56, 57, 62, 66, 75, 87, 90–93, 99, 106, 128, 142, 182, 185–189, 192, 194, 198, 216, 218–221, 239, 243, 244, 250, 252, 259
Cavalieri, B. (ca. 1598–1647) 32, 174, 239
Chebyshev, P. L. (1821–1894) 10, 83
Church, A. (b. 1903) 28, 62, 71, 239
Condorcet, M. J. A. N. (1743–1794) 48, 50, 55, 78, 239
Cooke, R. G. 21, 239
Copernicus, N. (1473–1543) 32

D'Alembert, J. L. (1717–1783) 44–47, 55, 182, 240
Daniell, P. J. (1889–1946) 116, 118, 201, 240
Dannemann, F. (1859–1936) 177, 240
Darboux, G. (1842–1917) 14, 65, 93–95, 102, 106, 190, 224, 228, 231, 240
Dedekind, R. (1831–1916) 11, 67–73, 75, 240
Dell'Angola, C. A. (1871–1956) 109, 240
De Morgan, A. (1806–1871) 70
Denjoy, A. (1884–1974) 14, 114, 193–195, 197, 198, 204, 233, 235, 240
Descartes, R. (1596–1650) 32–34, 36, 241, 258
Dickstein, S. (1851–1939) 45, 241
Dieudonné, J. (b. 1906) 22, 26, 50, 241
Dini, U. (1845–1918) 11, 12, 14, 65, 66, 95, 100–106, 108, 111, 157, 191, 213, 216, 229, 224–228, 239, 241
Diocles, (Second century B. C. E.) 30
Dirichlet, P. G. Lejeune (1805–1859) 41, 47, 49, 50, 53–55, 57, 60–63, 89, 94, 134, 144, 161, 189, 190, 192, 216, 220, 221, 241
Du Bois-Reymond, P. (1831–1889) 14, 65, 88, 93–95, 98, 103, 128, 133, 190, 192, 221, 224, 241
Dugac, P. (b. 1926) 91, 224, 242
Duhamel, J. M. C. (1797–1872) 88, 219
Dunford, N. (b. 1906) 196

Egorov, D. M. (1869–1931) 112, 113, 118, 119, 121, 124, 242, 247, 254
Einstein, A. (1879–1955) 247, 252
Eneström, G. (1852–1923) 47
Engels, F. (1820–1895) 178, 242
Euclid (ca. 365–300 B. C. E.) 30
Eudoxus of Cnidos (ca. 408–355 B. C. E.) 36, 170
Euler, L. (1707–1783) 34, 35, 40–48, 50, 54–57, 63, 64, 71, 78, 169, 180, 181, 185, 196, 237, 242, 244, 249, 252

Fatou, P. (1878–1929) 124
Felix, L. 233, 240
Fermat, P. (1601–1665) 36, 174–175
Fikhtengol'ts, G. M. (1888–1959) 121, 182, 242
Fischer, E. (1875–1959) 124, 137, 139–144, 146, 147, 148, 152, 242
Fomin, S. V. (1917–1975) 129, 247
Fourier, J. B. (1768–1830) 11, 40, 49–54, 59–62, 87, 88, 89, 124, 130–134, 138–140, 152, 185–186, 242, 244, 245, 246, 257
Fraser, C. G. 215, 242
Fréchet, M. (1878–1973) 19–21, 109, 114, 115, 138, 157, 163, 195–196, 199–205, 242
Fredholm, I. (1866–1927) 253
Frege, G. (1849–1925) 70–72, 237, 243
Freudenthal, H. (b. 1905) 57, 243
Freycinet, C. 219

Gaiduk, Yu. M. (b. 1914) 130, 243
Galilei, G. (1564–1642) 32–34, 175–177, 247, 255
Galois, E. (1811–1832) 191, 219, 220, 243
Garcet, H. 220
Gauss, K. F. (1777–1855) 91, 180, 182, 187, 221, 243
Gavurin, M. K. (b. 1912) 196
Gel'fand, I. M. (b. 1913) 196
Gerver, J. 222, 243
Glivenko, V. I. (1897–1940) 196, 205, 243
Gnedenko, B. V. (b. 1912) 23, 168, 243

Grabiner, J. V. 216, 243
Granger, G. G. (b. 1920) 48, 243
Grattan-Guinness, I. (b. 1941) 57, 63, 85, 151, 243
Graves, L. M. (b. 1896) 196
Gregory, J. (1638–1675) 32, 40, 41, 175, 176
Gyunter, N. M. (1871–1941) 76, 77, 185, 244, 256

Haar, A. (1885–1933) 83
Hadamard, J. (1865–1963) 19, 162, 197, 244
Hahn, H. (1879–1934) 98, 185, 244
Hake, H. 197, 244
Halmos, P. (b. 1916) 115, 116, 124, 129, 200, 244
Halphen, G. H. (1844–1888) 137, 244
Hankel, H. (1839–1873) 17, 22, 29, 42, 56, 62, 63, 65, 66, 72, 74, 75, 102, 158, 190, 222, 224, 244, 245
Hardy, G. H. (1877–1947) 14, 21, 66, 85–86, 99, 103, 151, 188, 222, 244
Harnack, A. (1851–1888) 107, 111, 116, 129–139, 149, 151, 164, 192, 243, 245
Hausdorff, F. (1868–1942) 164, 229, 245
Hawkins, T. (b. 1938) 50–53, 61, 63, 66, 106, 111, 116, 120, 132, 133, 138, 191, 193, 212, 214, 219, 245
Heine, H. E. (1821–1881) 87, 92–97, 111, 245
Hermite, Ch. (1822–1901) 233
Hayes, C. A. I. 201, 245
Hilbert, D. (1862–1943) 18, 23, 136, 138–140, 148, 243, 245
Hildebrandt, T. H. (b. 1888) 97–98, 245
Hippias (Fifth century B. C. E.) 30
Hobson, E. W. (1856–1933) 11, 12, 98, 101–104, 107, 109, 110, 124, 126, 128, 154, 229, 245
Hölder, O. (1859–1937) 147, 148, 152, 192
Hoppe, E. (1854–1928) 30, 171, 245
Houël, J. (1825–1886) 222, 245
Hugoniot, P. H. (1851–1887) 137, 246
Huytesbury, W. (ca. 1313–1372) 31

Ionescu-Tulcea, C. T. 196, 205, 246

Jordan, C. (1838–1922) 15, 22, 65, 66, 75, 76, 121, 133, 136, 137, 190, 246
Jourdain, P. E. B. (1879–1919) 223, 246

Kaczmarz, S. (1895–1939) 14, 150, 155, 246
Kantorovich, L. V. (1912–1986) 25, 156, 229, 246
Keldysh, L. V. (b. 1904) 166
Kempisty, S. (1892–1947) 167, 246
Kepler, J. (1571–1630) 173, 174
Keyser, C. J. (1862–1947) 79, 246
Khinchin, A. Ya. (1894–1959) 194, 197, 198, 246
Kitcher, P. (b. 1947) 44, 246
Kleene, S. C. (b. 1909) 25, 246
Klein, F. (1849–1925) 188, 246
Klement'ev, Z. I. (b. 1903) 210, 217, 246
Knopp, K. (1882–1957) 66, 227, 247
Kol'man, E. Ya. (1892–1977) 56, 57, 65, 247
Kolmogorov, A. N. (1903–1987) 23, 77, 124, 129, 177, 196, 198, 202–206, 208, 210, 247
König, J. (1849–1913) 191, 247
Königsberger, L. (1837–1921) 219
Korovkin, P. P. (b. 1913) 120, 247
Krasnosel'skii, M. A. (b. 1920) 152, 247
Kronecker, L. (1823–1891) 93, 120
Kudryavtsev, L. D. (b. 1923) 210, 247
Kuratowski, C. (1896–1980) 26, 27, 70, 73, 80, 228, 229, 247
Kuznetsov, B. G. (1903–1984) 177, 247

Lacroix, S. F. (1765–1843) 47–50, 56, 57, 59, 63, 65, 186, 187, 219, 247
Lagrange, J. L. (1736–1813) 35, 41, 44–48, 57, 62, 65, 182, 187, 214–218, 232, 241, 248, 252
Laguerre, E. (1834–1886) 83
Lamarle, A. H. E. (1806–1875) 219
Landau, E. (1877–1938) 121, 248
Laplace, P. S. (1749–1827) 55
Lavrent'ev, M. A. (b. 1900) 79, 248
Lebesgue, H. (1875–1941) 13, 14, 21, 56, 66, 74–77, 79, 97, 108, 110–114,

116–119, 121, 122, 129, 133, 136, 138–140, 142, 148, 147–152, 158, 159, 161–167, 183, 185, 190, 193–201, 205, 206, 216, 230, 237, 247, 248, 252, 254

Legendre, A. M. (1752–1833) 83, 183

Leibniz, G. W. (1646–1716) 29, 32, 39–41, 47, 66, 169, 177, 178–182, 187, 189, 196, 198–201, 249, 252

Liouville, J. (1809–1882) 83

Lipschitz, R. (1832–1903) 65

Littlewood, J. E. (1885–1957) 222

Lobachevskii, N. I. (1792–1856) 50, 57–62, 65, 192, 216, 220, 249

Looman, H. 197, 249

Luzin, N. N. (1883–1950) 62, 90, 105, 114, 125, 161–166, 181, 194, 225, 236, 249, 257

Lunts, G. L. (b. 1911) 58–60, 65, 249

Lyapunov, A. A. (1911–1973) 24, 237,

Lyapunov, A. M. (1857–1918) 136, 191

Mac Laurin, C. (1698–1746) 42, 43, 249

Maeda, F. 201, 249

Maistrov, L. E. (1920–1982) 23, 250

Mal'cev, A. I. (1909–1967) 26, 70, 250

Marcus, S. (b. 1925) 79, 250

Markov, A. A. (1856–1922) 191

Markushevich, A. I. (b. 1908) 35, 43, 45, 62, 180, 250

Mazurkiewicz, L. L. 189, 250

Mazurkiewicz, S. (1888–1945) 65, 167, 229–233, 250

Medvedev, F. A. (b. 1923) 22, 76, 111, 132, 160, 171, 191, 220, 250

Menger, K. 79, 250

Mengoli, R. (1625–1686) 40, 175, 235

Men'shov, D. E. (b. 1892) 24, 34, 114–116, 125, 127, 128, 229, 236, 250

Meyer, G. F. 188

Minkowski, H. (1864–1909) 152

Moigno, F. (1804–1864) 75, 99

Monge, G. (1746–1816) 45

Montel, R. (b. 1876) 109, 233, 240, 251

Moore, E. H. (1862–1932) 19, 23, 202, 251

Nalli, R. (1886–1965) 143–145, 251

Napier, J. (1550–1617) 32

Natanson, I. P. (1906–1964) 12, 73, 113, 122–123, 128, 141, 148, 158, 251

Natucci, A. 36, 47, 251

Newton, I. (1642–1727) 32, 34, 36–40, 47, 64, 66, 169, 177, 178–182, 187, 189, 196, 198–201, 236, 251

Nicomedes, (Second century B. C. E.) 30

Nikliborc, L. 155, 246

Nikodým, O. (b. 1887) 198–201, 207, 251

Nikol'skii, S. M. (b. 1905) 210, 247, 251

Noaillion, P. 155

Oldenburg, H. (1615?–1677) 251

Oresme, N. (ca. 1323–1382) 31, 172

Orlicz, W. (1903–1990) 152–156, 231, 232, 237, 249, 251

Osgood, W. F. (1864–1943) 98, 108, 112, 237, 252

Ostrogradskii, M. V. (1801–1861) 190, 252

Oxtoby, J. C. 229, 231, 252

Paplauskas, A. B. (b. 1931) 22, 45, 46, 63, 88, 92, 136, 145, 184, 185, 252

Pappus of Alexandria (Third and Fourth centuries) 30, 170, 173, 250

Pascal, E. (1865–1940) 65, 212, 216, 252

Pasch, M. (1843–1930) 65, 66, 75, 212, 216, 219, 252

Pauc, C. Y. 201, 245

Peano, G. (1858–1932) 22, 66, 70, 71, 72, 75, 76, 121, 133, 136, 137, 182, 190, 191, 220, 250, 252

Peirce, C. S. (1839–1914) 70

Perrin, L. 234

Perron, O. (b. 1880) 14, 194, 195, 197, 237, 249, 252

Pesin, I. N. (b. 1930) 137, 193–194, 252

Pettis, B. J. (b. 1913) 196

Phillips, R. S. (b. 1913) 196

Picone, M. (b. 1885) 23

Pincherle, S. (1853–1936) 21, 97, 107–109, 142, 145, 149, 252

Plancherel, M. 142, 252

Pogrebysskii, I. B. (1906–1971) 179, 184, 252
Poinsot, L. (1777–1859) 219
Poisson, S. D. (1781–1840) 45, 140, 182, 183, 186, 187
Price, G. B. (b. 1905) 196
Pringsheim, A. (1850–1941) 85, 91, 98, 252
Privalov, I. I. (1891–1941) 145, 252
Ptolemy, Cladius (d. ca. 170) 30, 172
Pythagoras (Sixth century B. C. E.) 177

Raabe, J. L. (1801–1859) 219, 253
Rademacher, H. (1892–1969) 83
Radon, J. (1877–1956) 76, 116–117, 142, 151, 195, 198–201, 207, 253
Remez, E. Ya. (b. 1896) 190, 253
Reyneaux, Ch. R. (1656–1728) 41, 253
Rickart, C. E. (b. 1913) 196
Riemann, B. (1826–1866) 13, 14, 15, 16, 20, 22, 62, 63, 65, 84, 95, 106, 107, 112, 114, 121, 130–134, 137, 138, 139, 150, 164, 189–190, 198, 221–224, 240, 243, 248, 253
Riesz, F. (1880–1956) 98, 114, 115, 118, 119, 121–126, 136, 139–151, 154, 155, 194, 197, 253
Ringenberg, L. A. 202, 254
Roberval, G. P. (1602–1675) 34, 174, 175
Rosenthal, A. (1887–1959) 63, 85, 98, 106, 110, 138, 229, 254
Rozhanskaya, M. M. (b. 1928) 31, 172, 254
Rubini, R. 219
Russell, B. (1872–1970) 70, 73, 74, 162, 254
Rutickii, Ya. B. (b. 1922) 152, 247
Rybnikov, K. A. (b. 1913) 49, 254

Saks, S. (1897–1942) 14, 21, 200, 229–233, 254
Schauder, J. S. (1889–1940) 238
Scheffer, L. (1859–1885) 192
Schmidt, E. (1876–1959) 139, 142, 254
Schönflies, A. (1853–1928) 192, 229, 254
Schramm, M. (b. 1928) 30, 172, 255
Schröder, E. (1841–1902) 70, 72–74, 255

Schwarz, H. A. (1843–1921) 223, 224, 255
Scriba, Ch. J. (b. 1959) 176, 255
Seidel, Ph. L. (1821–1896) 88–93, 95, 100, 255
Serret, J. A. (1819–1885) 225
Severini, C. (1872–1951) 113, 145, 255
Shabat, B. V. (1917–1987) 78, 248
Shatunovskii, S. O. (1859–1929) 23, 255
Shchegol'kov, E. A. (b. 1917) 164, 255
Shikhanovich, Yu. A. (b. 1933) 26, 70, 255
Shilov, G. E. (1917–1975) 22, 255
Sierpiński, W. (1882–1969) 161, 167, 249, 255
Sikorski, R. 205, 255
Sirvint, Yu. F. (1913–1942) 109, 111, 255
Skvortsov, V. A. (b. 1935) 198, 209, 257
Smirnov, V. I. (1887–1974) 185, 209, 256
Smith, H. J. S. (1827–1883) 17, 66, 106, 190, 256
Smith, H. L. (1892–1950) 23, 203, 251
Smolyanov, O. G. (b. 1938) 196, 236
Sobolev, S. L. (b. 1908) 185, 209, 210, 256
Steiner, H. G. 61, 69, 72, 256
Steinhaus, H. (1887–1972) 14, 150, 229, 231, 246
Steklov, V. A. (1864–1926) 136, 145, 252, 256
Stieltjes, T. J. (1856–1894) 12, 17, 21, 116–119, 150, 151, 155, 191, 195, 197, 199, 201, 233, 240, 243, 244, 248, 256
Stokes, G. G. (1819–1903) 57, 62, 88–91, 99–101, 244, 256
Stolz, O. (1842–1905) 93, 192, 256
Sturm, J. Ch. F. (1803–1855) 83
Styazhkin, N. I. (1932–1986) 70, 256
Suslin, M. Ya. (1894–1919) 166
Swineshead, R. (fl. ca. 1350) 31, 172

Talalyan, A. A. (b. 1928) 127, 128, 238, 256
Tarski, A. (b. 1902) 27, 72, 256
Taylor, B. (1685–1731) 41, 44, 215, 218, 235

Thābit ibn-Qurra, (826–901) 171
Thomae, K. J. (1840–1921) 95, 106, 190, 192, 241, 256
Thomé, L. W. (1841–1910) 91, 92, 94, 256
Timchenko, I. Yu. (1862–1939) 43, 45, 46, 54, 55, 191, 257
Tolstov, G. P. (b. 1911) 24, 130, 131, 136, 237, 257
Torricelli, E. (1608–1647) 34, 176

Ul'yanov, P. L. (b. 1923) 128, 155, 257
Uspenskii, V. A. (b. 1930) 161, 257

Valerio, L. (1552–1618) 175
Vallée-Poussin, Ch. (1866–1962) 16, 76, 117, 118, 164, 192, 244, 257
Varignon, P. (1654–1722) 253
Vasilescu, F. (1897–1958) 79
Vinogradova, I. A. 198, 204, 209, 257
Vitali, G. (1875–1932) 152, 194, 201, 245, 252, 257
Volterra, V. (1860–1940) 19, 21, 102, 106, 190, 192, 257
Voronoi, G. F. (1868–1908) 191

Wallis, J. (1616–1703) 40, 175
Weierstrass, K. (1815–1897) 65, 78, 79, 88, 91–92, 96–97, 109, 219, 221–224, 227, 231, 242, 244, 257
Wells, H. G. (1866–1946) 10
Weyl, H. (1885–1955) 114, 115, 122, 142, 258
Wieleitner, H. (1874–1931) 88, 258
Wiener, N. (1894–1964) 234, 258

Young, G. C. (1868–1944) 151, 244
Young, R. C. (b. 1900) 80, 258
Young, W. H. (1863–1942) 97–98, 116, 118, 121, 151–156, 159, 167, 191, 195, 201, 244, 258
Yushkevich, A. P. (b. 1906) 29–33, 35, 36, 38, 40–42, 44, 47, 48, 55–57, 171, 183, 187, 258

Zeller, K. (b. 1924) 85, 258
Zermelo, E. (1871–1953) 87, 162, 258
Zeuthen, H. G. (1839–1920) 29, 175, 176, 179, 258
Zubov, V. P. (1899–1963) 177, 258